IMAGES
of America

LOVE CANAL

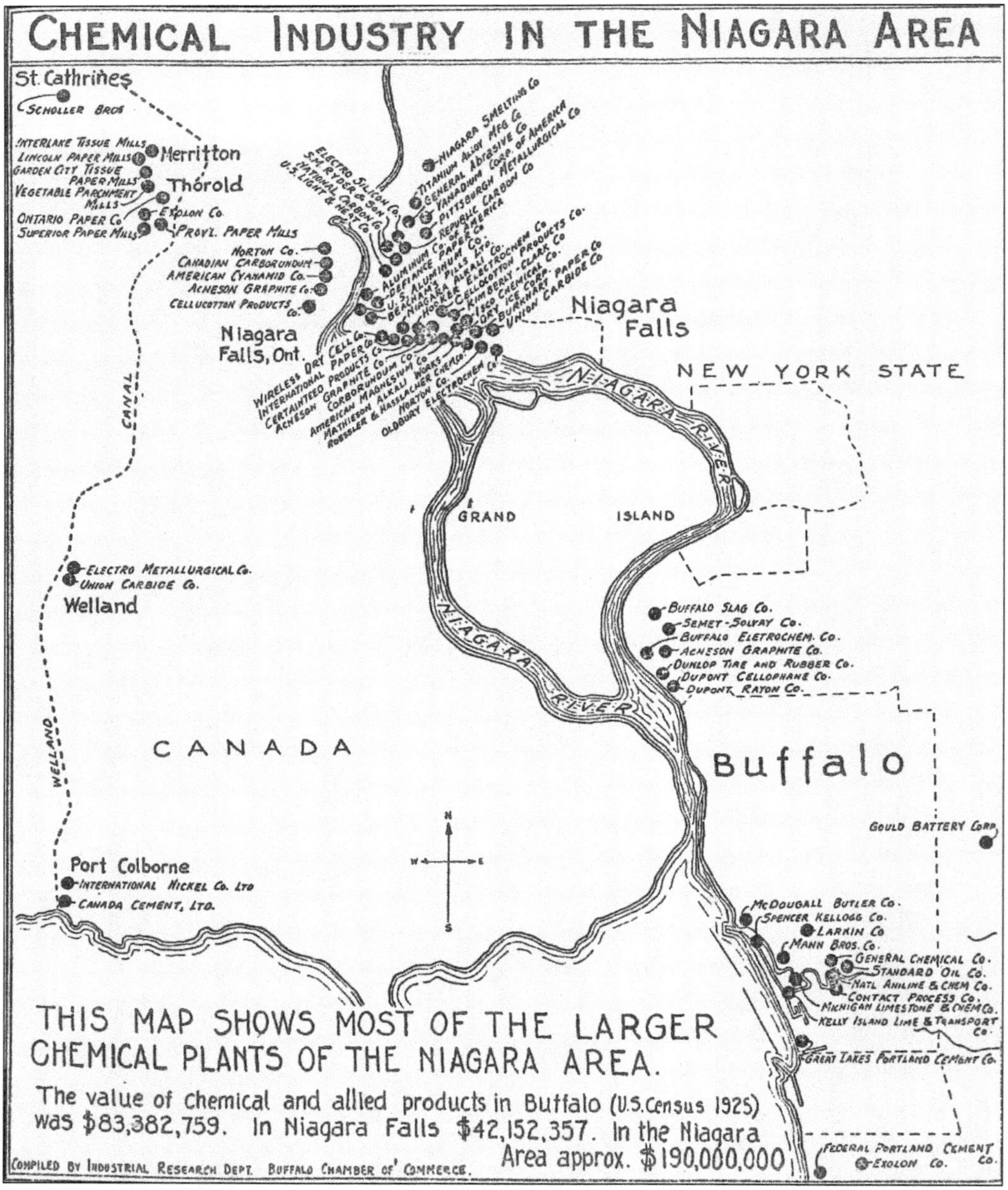

A 1925 map of the chemical industry in the Niagara area shows the dense occupation of the Niagara River shoreline in Niagara Falls, New York, by chemical plants, including the "Hooker Electrochem Co." (Buffalo Chamber of Commerce.)

On the Cover: Folks turn the first sod for William T. Love's power canal and model city project on May 23, 1894. (Photograph by J.C. Hooker of Lewiston, New York; courtesy of the History Center of Niagara County.)

IMAGES
of America

LOVE CANAL

Penelope Ploughman, PhD, JD

ARCADIA
PUBLISHING

ISBN 978-0-7385-7560-5

Published by Arcadia Publishing
Charleston, South Carolina

Library of Congress Control Number: 2010943010

For all general information, please contact Arcadia Publishing:
Telephone 843-853-2070
Fax 843-853-0044
E-mail sales@arcadiapublishing.com
For customer service and orders:
Toll-Free 1-888-313-2665

Visit us on the Internet at www.arcadiapublishing.com

To the archivists and librarians of the past, present, and future, without whom history would not be found, recorded, or preserved

Contents

Acknowledgments

I am indebted to John Edens and William Offhaus of the State University of New York at Buffalo, Love Canal Archives, for their advice and assistance in preserving and digitizing my Love Canal photographs and for making them available on the Western New York Legacy Web site. In addition, I thank Ann Marie Linnabery of the History Center of Niagara County; Linda Reinumagi of the Niagara Falls Public Library, Local History Department; and Cynthia Van Ness of the Buffalo and Erie County Historical Society for their help in finding and scanning many of the unique historical photographs included in this book. I also thank Frank Cornell, former director of the Love Canal Area Revitalization Agency, for his time and advice and for engraving the history of Love Canal on a stone monument and placing it near the site so we never forget what happened at Love Canal. I am immensely grateful to Warren Philipson, PhD, for allowing me to include his unique and irreplaceable collection of aerial photographs. Last, but certainly not least, I thank my editors at Arcadia Publishing, Caitrin Cunningham, Erin Roche, and Rebekah Mower, for their guidance, patience, and encouragement.

Some of the images in this volume appear courtesy of the State University of New York at Buffalo, Love Canal Archives (SUNY); the New York State Department of Health (DOH); the History Center of Niagara County (HC); the Niagara Falls Public Library (NFPL); the Warren Philipson Collection (WPC); the Buffalo Chamber of Commerce (BCC); Edward Dean Adams's *Niagara Power* (Adams); or the Love Canal Area Revitalization Agency (LCARA). Unless otherwise noted, all photographs in this volume were taken by the author.

INTRODUCTION

This is the story of Love Canal from its beginning as part of a power canal and model city dream through its infamy as an environmental nightmare. The Love Canal area of the city of Niagara Falls, New York, got its name in 1894 when it was selected as the location for a canal that would provide electrical power for a new model city envisioned by entrepreneur William T. Love. Love's model city was to be built in a rural area outside of the town of Lewiston, New York, and the power canal was to be used to generate hydroelectric power. The power was to be given away free of charge—in unlimited amounts and for 25 years—to manufacturers and industries willing to locate there. The canal was the focal point, primary resource, and major attraction for the development of the city. The canal was designed to provide power, water, and transportation for industry. Due to the massive and awesome nature of Niagara Falls, the potential for the generation of hydroelectric power was limited only by the technology in existence at the time. Because the technology of the day was direct current hydroelectric transmission, properties located closest to the generating facility were highly sought after by industry and manufacturers. Although power and ship canal construction projects and company and model towns had been designed and constructed elsewhere, what made Love's dream so significant was the sheer scale of the project he envisioned: a seven-mile-long canal dug from the Niagara River to the brink of the Niagara escarpment, from where the diverted water would plummet 210 feet to a generating station and then flow on to the brink of the lower Niagara River, where it would again drop another 80 feet to a second generating station below before finally entering the lower Niagara River. The power canal and the massive, beautiful model city Love designed were unlike anything ever seen before. Love selected the Niagara area due to the many advantages it offered: abundant natural resources, railroads, waterways, abundant water supply, attractive scenery, productive country, and good climate. Referring to the location, Love said, "It is par excellence, the location of the world for such an enterprise as ours." At the time, the technology existed to dig power and shipping canals, such as that proposed by Love—in fact, canals could have been dug that were large enough to so significantly curtail the flow of the Niagara River over Niagara Falls as to shut it off. Before 1885, there were no laws protecting Niagara Falls, and until 1906, there were no laws prohibiting or regulating the digging of such power canals or regulating the extraordinary extent of water diversion from the Niagara River that was possible with the technology of the day.

In 1895, Nikola Tesla's discovery of alternating electrical current and a national economic downturn ended Love's dream. In 1906, any further dreams of power canals were ended forever when the US Congress passed the Burton Act, which prohibited any additional diversion of water from the Niagara River in order to protect the falls. Niagara's power and water, combined with the abundant local natural resources and the developed railroad and water transportation systems, made Niagara Falls "the greatest electro-chemical and electro-metallurgical center in the world," according to the Buffalo Chamber of Commerce. The development of Niagara hydroelectric power also marked the beginning of large-scale electrochemical and electrometallurgical manufacturing in the Niagara area. One of the first companies was the Hooker Electrochemical Company. Electrochemical and electrometallurgical industries depend upon a continuous and abundant supply of electricity and electrothermic or electrolytic processes for production of materials and products. Electrothermic processes use high-temperature electrical furnaces to combine two or more natural elements to make a new product. For example, the abrasive carborundum is a result of the heat fusion of carbon and silica. Electrolytic processes use electricity to break down natural elements in order to produce a new material. For example, running electrical current through ordinary salt produces chlorine and caustic soda (lye). Niagara's power was used by large-scale factories to make products that quickly became part of modern life, including aluminum, abrasives, ferro-alloys (special metals like titanium, chromium, and tungsten), sodium, silicon, magnesium, potassium, phosphorous, calcium carbide, graphite, chlorine, and chemicals such as paraformaldehyde, trioxymethylene, methyl alcohol, methyl acetone, hydrochloric acid, carbon tetrachloride, sulfur chloride, caustic potash, phosphorus compounds, and cyanamid.

By 1925, Niagara Falls was the electrochemical and electrometallurgical center of the world. According to the 1925 Census of Manufacturers, there were 116 industrial establishments in Niagara Falls—21 of which were chemical plants. These 21 chemical plants employed 44 percent of the total industrial workforce of Niagara Falls. During World War II, the Niagara area chemical industry, including the Hooker Electrochemical Company, provided many essential war-related products, including the following: smoke pots, flares (hexachlorobenzene); disinfectants and decontaminants (bleaching powders and chlorine); sodium sulfides, sodium sulfhydrates, and sodium tetrasulfide for the leather industries; caustics, aluminum chloride, chlorinated paraffin, and organic sulfides for the lubricant industries; synthetic rubbers; poisons and poisonous gases (chlorine, thionyl chloride, and arsenic trichloride); magnesium; and chemicals used in medicines, dyes, plastics, water repellants, soaps, and rayon and other synthetic textiles. Hooker Electrochemical alone produced 50 percent of the synthetic rubber and more than 2.5 billion pounds of chemical products during the war years (1940–1945). Hooker also participated in the Manhattan Project, which helped develop the atomic bomb, and operated a plant for the Atomic Energy Commission in Model City, New York—a town that developed where William T. Love had envisioned his dream city. Hooker acquired Love's old canal property in 1942, and from 1942 to 1952, the company disposed of over 21,000 tons of chemical wastes in the canal. By 1978, some 82 years after Love's dream of a power canal and model city ended, the remnant of his canal became part of an American Dream suburban neighborhood of single-family homes surrounding an elementary school. The neighborhood would be the location of the infamous Love Canal toxic waste landfill disaster and the harbinger of the worldwide hazardous waste disposal crisis.

One

Harnessing Niagara The Beginning

The history of Love Canal is inseparable from the history of the mighty falls and the revolutionary quest to harness their power. Niagara Falls became known as the electrical power and the electrochemical capital of the world. In 1853, Caleb Woodhull wrote, "Here is a power almost illimitable; constantly wasted, yet never diminished—constantly exerted, yet never exhausted—gazed upon, admired, wondered at, but never hitherto controlled." (NFPL.)

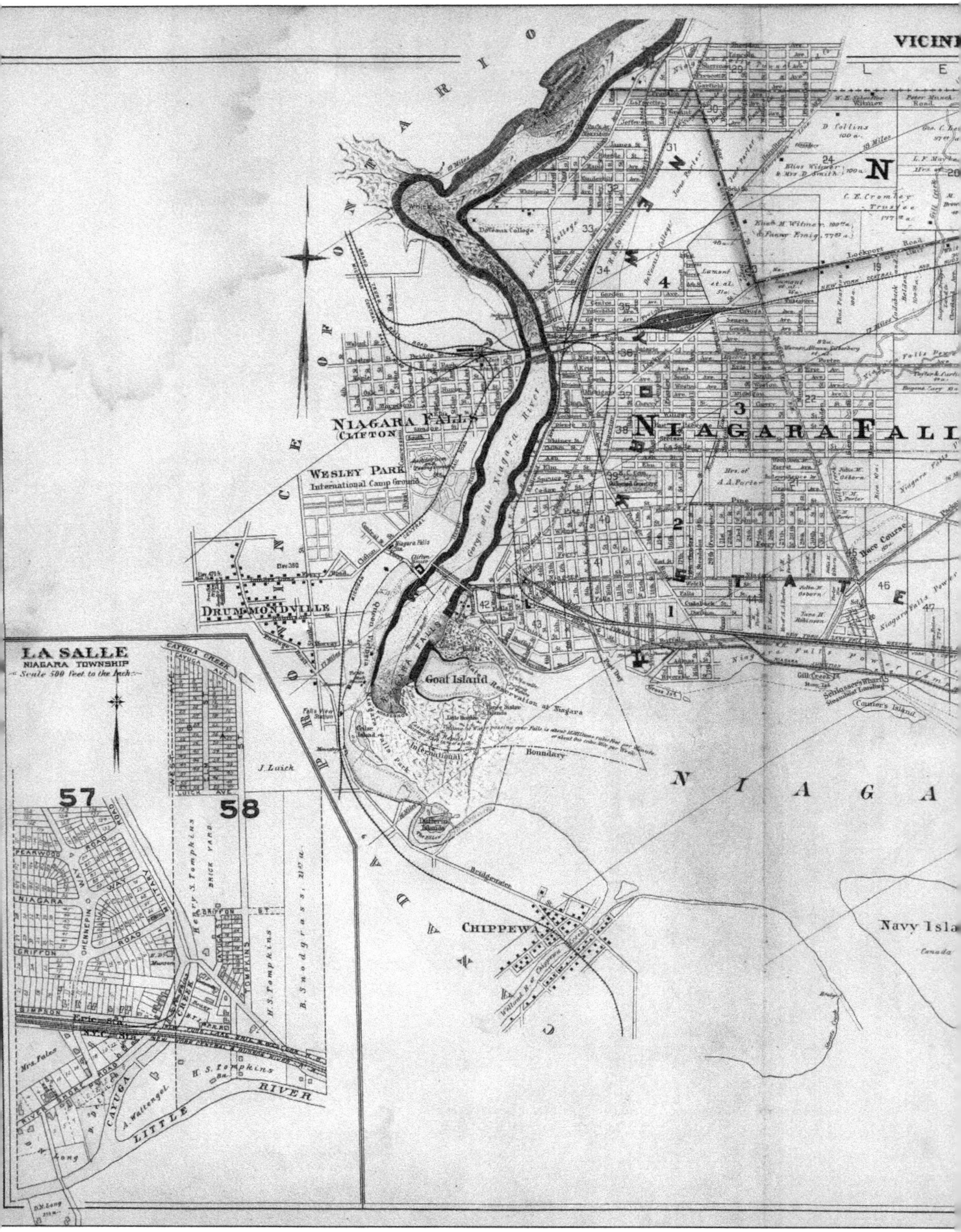
VICINI
NIAGARA FALLS
CLIFTON
WESLEY PARK
International Camp Ground
DRUMMONDVILLE
LA SALLE
NIAGARA TOWNSHIP
Scale 500 feet to the Inch
57
58
Goat Island
International
Boundary
CHIPPEWA
Navy Isla
Canada
Gorge of the Niagara River
Lockport Road
Race Course
Niagara Falls Park
Dufferin Islands
Cayuga Island
Schlosser's Wharf
Steamboat Landing
Henry S. Tompkins
BRICK YARD
H. S. Tompkins
B. Snodgrass
J. Laick
Mrs. Fales
A. Wattengel
LITTLE RIVER
CAYUGA CREEK
NIAGARA
GRIFFON
PEARWOOD
HENNEPIN
MILITARY
Erie

This 1893 map of Niagara Falls shows the state reservation at Niagara Falls; the Queen Victoria Niagara Falls Park; Goat Island; Port Day; Grass Island; the lands of the Niagara Power Company; the lands of the Niagara Falls Power Company; and the village of LaSalle, including Lot 60, owned by David Long, where Love's canal was started in May 1894. The first known use of Niagara River water for power was in 1759, when Daniel Joncaire dug a small canal just above the American Falls to power his sawmill. At this time, the power of flowing water was directly and mechanically harnessed through a waterwheel. In 1805, Augustus and Peter Porter purchased this canal and the American Falls from New York State and widened the canal to provide hydraulic power for their gristmill and tannery.

During the 19th and early 20th centuries, the falls were often called "the sublimest of nature's works." Since they were scenic, exciting, and accessible through New York's turnpike and the Erie Canal, the falls were initially a premier tourist destination. Niagara Falls would also become the "honeymoon capital of the world" due, in part, to the arousing and erotic effects the falls allegedly had on newlyweds. By 1930, sixty percent of the US population and 80 percent of Canada's were within one day's ride of Niagara Falls. (BCC.)

The falls became the ultimate challenge, a chance to conquer nature, and a mecca for daredevils of all sorts seeking to survive either going over the cataract in a barrel or by tightrope walking above it. Maria Spelterini, the only woman to tightrope walk across the gorge, crosses wearing baskets on her feet in 1876. (NFPL.)

Annie Edison Taylor, shown here with her barrel, went over the falls on October 24, 1901. Private ownership of the shoreline, the Niagara River islands, and the American Falls resulted in tourist traps where people paid for a glimpse of the falls and the daredevils who challenged them. By the late 1860s, many who feared the spoiling of the falls' grandeur became preservationists and started the "Free Niagara" movement. (NFPL.)

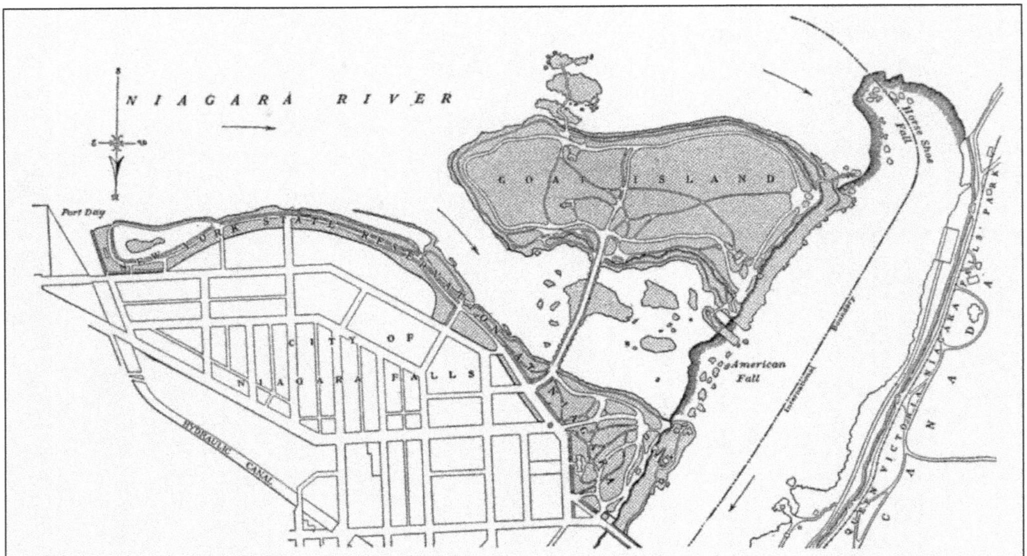

STATE RESERVATION AT NIAGARA
From Dow's "Anthropology and Bibliography of Niagara Falls"

On March 14, 1883, Gov. Grover Cleveland signed a bill providing for the purchase of land and the appointment of a reservation board. The land was a reservation rather than a park because a park meant tidy flower beds, lawns, benches, and sidewalks, whereas this land was to be kept as close as possible to the natural environment. This 1885 map shows the dimensions of the reservation. (Adams.)

This 1906 photograph shows a primary vantage point of the 412-acre reservation (300 acres of which were under water). The reservation included vantage points from where the falls could be seen. The land had been owned by 25 property owners, most of whom had developed their property and deliberately fenced off the non-paying public's view of the falls. (NFPL.)

The most sublime of nature's works became the ultimate technological challenge of the day: harnessing the Niagara and producing usable power. The attitude was that nature was there to be conquered, and natural resources were there for the taking. In less than a generation, the technology of waterpower advanced from waterwheels, to hydroelectric direct current, to hydroelectric alternating current, to transmission of alternating current by wire. (NFPL.)

Edward Dean Adams (1846–1931) wrote, "The falls of Niagara were successively in the possession of the Indians by inheritance, the French by discovery, the English by conquest, the American colonists by revolution, and the State of New York by cession, treaty and purchase." Adams also said that the story of the falls in the 20th century had been one of efforts to control them—efforts to control development that threatened the beauty and scenic wonder of the falls and efforts to control and harness the power of the falls. (NFPL.)

About 16 years after the quest to Free Niagara started, the reservation was opened on July 15, 1885. On May 24, 1888, the Canadian government opened Queen Victoria Niagara Falls Park. Both governments began balancing private and public interests in the falls (profit versus preservation) and managing the resources of the falls, like their scenic beauty, seemingly unlimited water and power, and the demands of tourism, manufacturing, and industry. (NFPL.)

This diagram demonstrates the distribution of Niagara power. The Niagara River and the falls would supply the fuel for the engine of one the greatest workshops of the world and start a second industrial revolution. This second industrial revolution was based upon the availability and control of electricity and resulted in many major electrical inventions, including the following: the electric motor (1821), the electromagnet (1823), the electromagnetic telegraph (1832), the incandescent light bulb (1835), the underwater telegraph cable (1858), the direct current motor (1873), the electric streetcar (1874), the dynamo (1875), the telephone (1876), the induction motor (1877), the photovoltaic cell (1883), the two-phase alternating current induction motor (1883), the alternating current transformer (1885), the polyphase alternating current power system (1888), the photoelectric cell (1888), the tungsten filament (1890), the radio (1893), the radiotelegraph (1895), and the development of both electrochemistry and electrometallurgy. (Adams.)

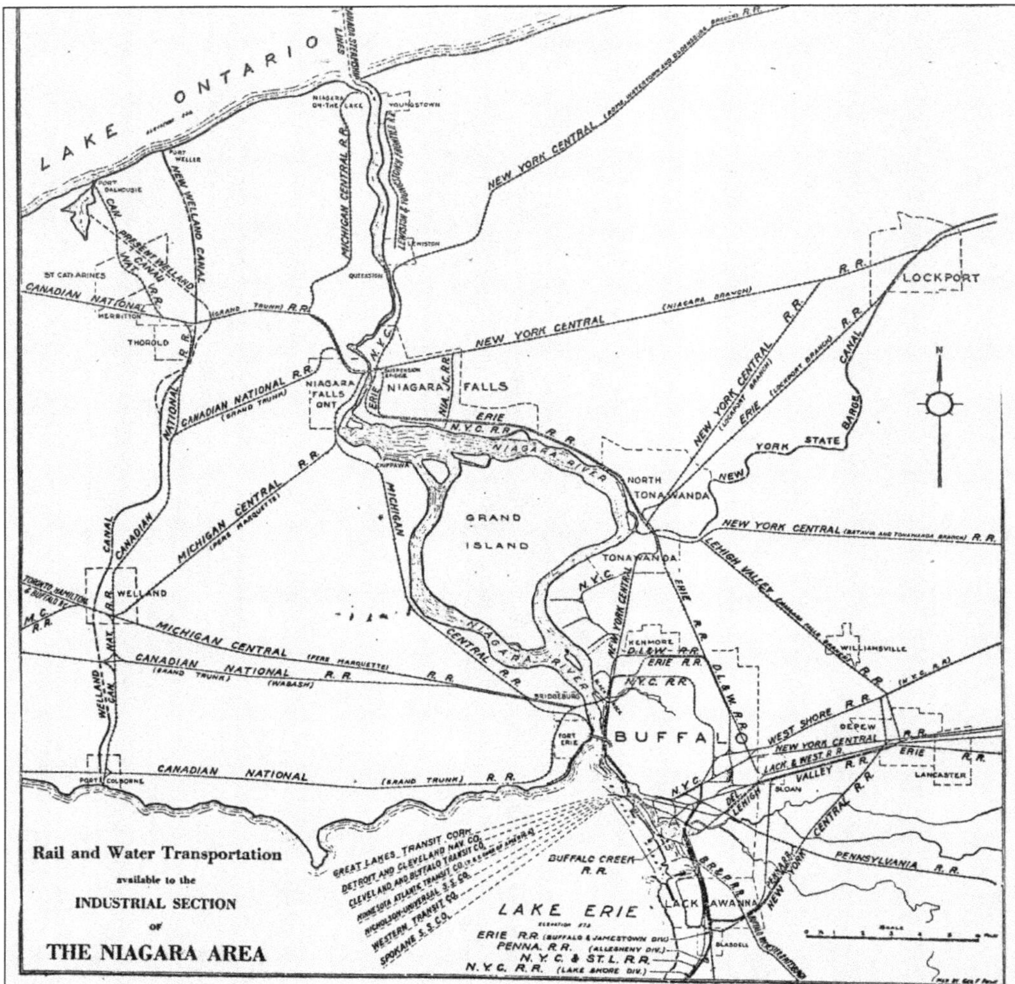

INBOUND COMMODITIES		OUTBOUND COMMODITIES	
via Lake	via Barge Canal	via Lake	via Barge Canal
AUTOMOBILES	SULPHUR	COAL	GRAIN & GRAIN PRODUCTS
ORE	LUMBER	CEMENT	PIG IRON
LIMESTONE	SALT	IRON & STEEL	IRON & STEEL
COAL	OIL	PACKAGE FREIGHT	CEMENT
SAND & GRAVEL	SUGAR	SUGAR	FLAXSEED
GRAIN	FLAXSEED	OIL, FUEL	CHEMICALS
CEMENT	FLUORSPAR	PIG IRON	OIL
PACKAGE FREIGHT	PHOSPATE ROCK	AUTOMOBILES	
OIL, FUEL			

Beginning in the 19th century, the technology of railway and canal engineering and shipping rapidly advanced. The 4-foot-deep, 40-foot-wide, 363-mile-long Erie Canal was started in 1817 and completed in 1825—it was quite a feat. With 40 wooden locks, the 27-mile-long Canadian Welland shipping canal was started in 1824 and completed in 1829. In 1836, Capt. W.G. Williams, a government engineer, surveyed Niagara County to determine the feasibility of constructing an all-American canal that would compete for shipping with the Welland Canal. Williams noted that while Lewiston, located at the bottom of the nearly 300-foot Niagara escarpment, made the construction of a shipping canal prohibitively difficult, it offered an outstanding location for a massive waterpower station. (BCC.)

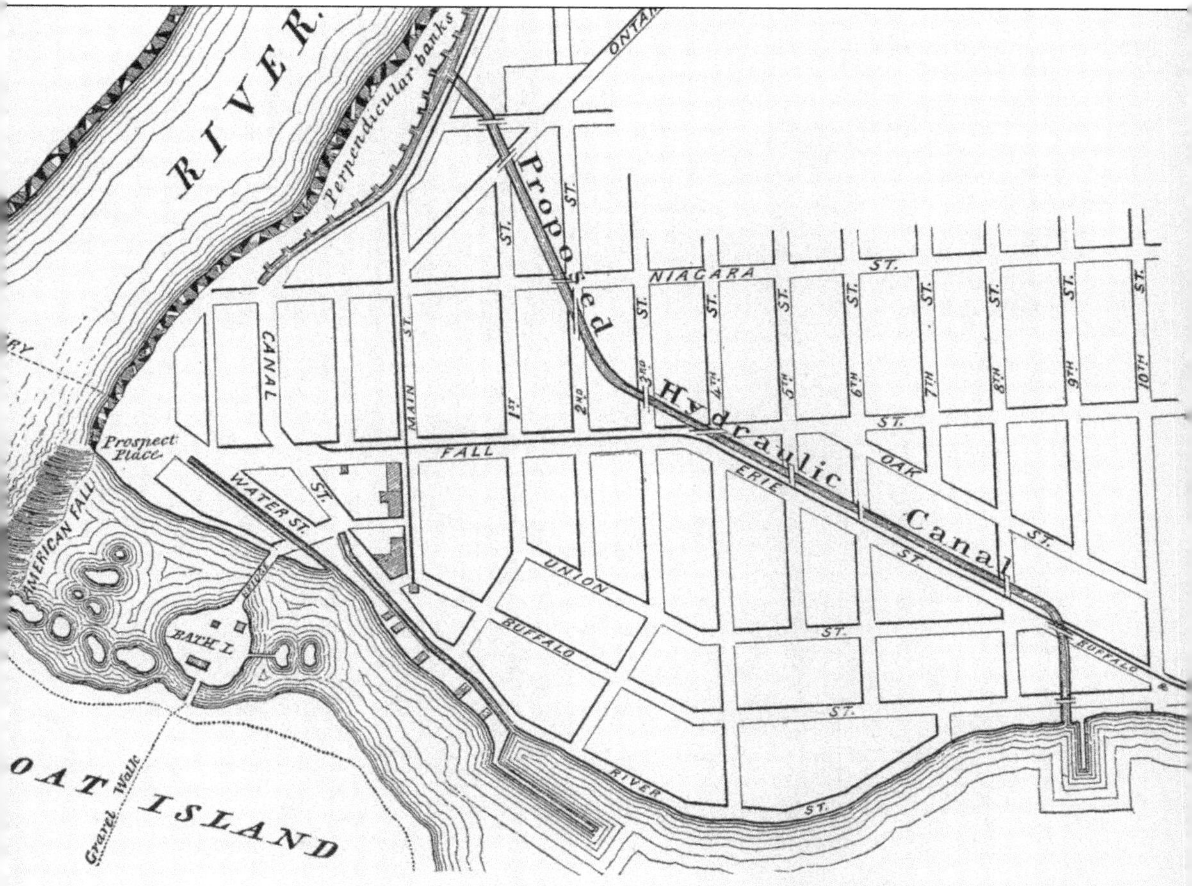

MAP OF NIAGARA FALLS AND VILLAGE

By P. Emslie, December, 1846

In 1825, Augustus Porter issued an invitation to anyone interested in developing the waterpower at Niagara Falls. This was the first advertisement seeking capital for power. In 1847, Porter, who owned the American Falls and much of the Niagara River shoreline, offered a canal right-of-way "to any interested Capitalists and Manufacturers" who would undertake the construction of a hydraulic canal. The canal was to begin on the river a mile above the falls and run across town to the riverbank a quarter of a mile below the falls. A basin one mile in length along the riverbank, would receive and discharge the water from the canal through raceways to each factory site with a fall of up to 190 feet. In 1853, the first attempt at a hydraulic canal was begun by the Niagara Falls Hydraulic Company, founded by Caleb Woodhull. The plan was to run factories with waterpower drawn from the upper Niagara River and to construct a navigable hydraulic canal with gates, wharves, and bridges. (Adams.)

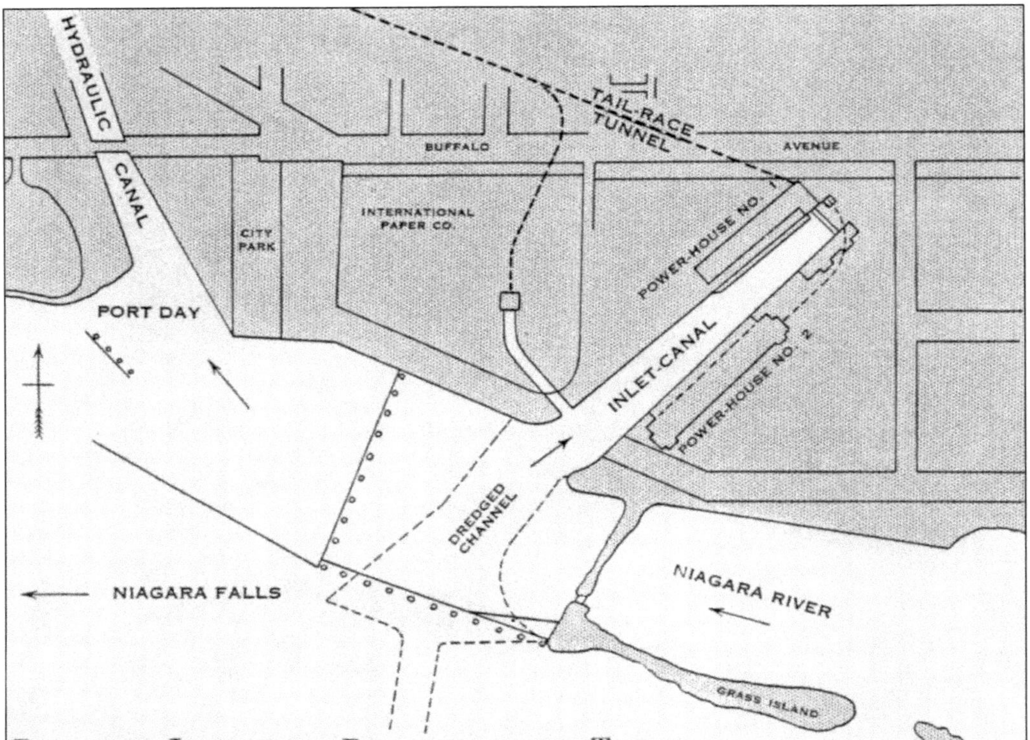

Diagram of Inlet-canal, Power-houses, and Tunnel

By July 4, 1857, the entrance at Port Day and river entrance sections of the hydraulic canal were completed and opened to navigation. All electricity prior to this point in time was direct current, which could not transmit electricity beyond the immediate area of the power station. This restriction necessitated that all powered manufacturing was tethered to the generating station. Icebreakers were used to keep the canal open in winter (Above, Adams; left, NFPL.)

In 1856, the Niagara Falls Hydraulic Company, having underestimated the construction costs and having insufficient capital to continue the project, changed its name to the Niagara Falls Water Power Company and acquired additional capital. This 1857 photograph shows water diverted from the upper Niagara River falling from a tailrace on the High Bank. No factory was in operation at the time. (NFPL.)

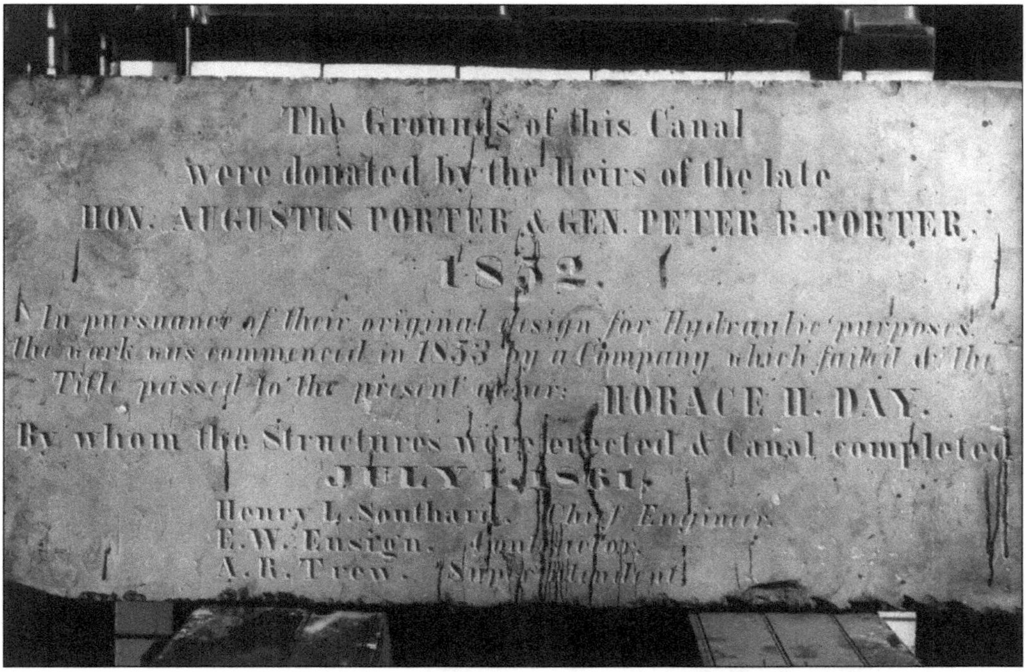

In 1860, the Niagara Falls Water Power Company was purchased by Horace H. Day and became known as the Niagara Falls Canal Company. A narrow canal was competed in 1861, and a dedication tablet was engraved and placed at the site. (HC.)

This 1875 photograph shows the Gaskill Flour Mill on the High Bank. This was the first mill built on the site. It used only an eighth of the potential power available and only 25 feet of the 210-foot head available for waterfall. By 1876, a one-mile-long canal, with a capacity of about 27,000 horsepower and cut through solid rock, had been completed. (NFPL.)

On May 1, 1877, Jacob F. Schoelkopf purchased the Niagara Falls Canal Company at foreclosure for $76,000. The purchase included the inlet from the Niagara River, an unfinished canal (60 feet wide and 11 feet deep at the inlet and an average 22 feet wide and 10 feet deep), the riparian rights, and 45 acres of land on the High Bank where the manufacturers would build. Under his leadership, the Niagara Falls Canal Company made the hydraulic canal commercially profitable, and over 100 mill and factory sites and 300 cottage lots were made available for purchase. In 1956, the canal was drained and the successor company, the Niagara Mohawk Power Company, gave the land to the city, which used it as a landfill. (HC.)

In 1878, Jacob Schoelkopf formed a new company called the Niagara Falls Hydraulic Power and Manufacturing Company and acquired, improved, and enlarged the hydraulic canal, as seen in these photographs. The Schoelkopf & Matthews Flour Mill was erected and utilized a 50-foot head of water that was discharged down the cliff 150 feet to the river below. In 1881, the company built the first hydroelectric power station to supply direct current electricity for commercial purposes on the site of the Quigley Pulp Mill. In 1886, Charles B. Gaskill founded the Niagara River Hydraulic Tunnel, Power, and Sewer Company. This company received a special charter from New York State giving it the authority to take an unlimited amount of water from the Niagara River for power generation of 200,000 horsepower. (Both, HC.)

In 1889, the Gaskill Company became the Niagara Falls Power Company, and the Cataract Construction Company was formed as its financial agent and charged with finding the best method of developing power. Cataract determined that a system for electrical power transmission with a single generating facility and a tailrace discharge tunnel was the most efficient and economical. The tunnel took three years to complete and cost 28 lives. It was 18 feet by 21 feet and 7,000 feet long and lined with 16 million bricks. In the 1970s, the tailrace was put to new use as an outfall for treated wastewater ("greywater") when the powerhouses were demolished to make way for a sewage treatment plant. Unfortunately, the old wheel pits, which could have been reengineered to generate power from the wastewater, were backfilled with demolition debris. (Both, HC.)

This 1898 photograph of the High Bank shows more than a dozen tailraces from mills and factories fed by the canal. Firms included, from left to right, the Central Milling Company; the Niagara Wood Paper Company; Schoelkopf & Matthews Flour Mill; Pettebone Pulp Mill; Gaskill Flouring Mill; Niagara Falls City Water Works; The Pittsburgh Reduction Company's aluminum manufacturing mill; Cliff Paper Company (upper mill); Cliff Paper Mill (lower mill); and the Oneida Community Mill. The Niagara Falls Hydraulic & Manufacturing Company's Power Station No. 2 is seen to the right at the bottom of the gorge (the second structure from the right with signage for the Pittsburgh Reduction Company on it). Industrialists, scientists, and inventors coveted each other's progress and continually dreamed up new inventions and improvements on the power technology of the day. Many dreamed of ending the restrictions that direct current had on the expansion of industry, and one succeeded—Nikola Tesla. (HC.)

In 1882, Nikola Tesla, the 26-year-old son of a Greek Orthodox priest from Croatia, discovered the principle of a rotating magnetic field produced by two asynchronous alternating currents. Tesla's inventions include a motor utilizing alternating current electricity, the Tesla coil (still used today), and more than 40 patented alternating current mechanisms. Tesla is credited with single-handedly changing the course of science and technology and paving the way for the exponential increases in power generation. In 1893, Tesla developed a polyphase (three-phase) system of alternating power transmission by wire. This was one of the major technological developments of the 19th century. In 1894, massive 29-ton turbines and Tesla's polyphase motors were lowered into the wheel pits of the powerhouse of the Niagara Falls Power Company, and in 1895, the City of Niagara Falls was lit by electricity for the first time. (HC.)

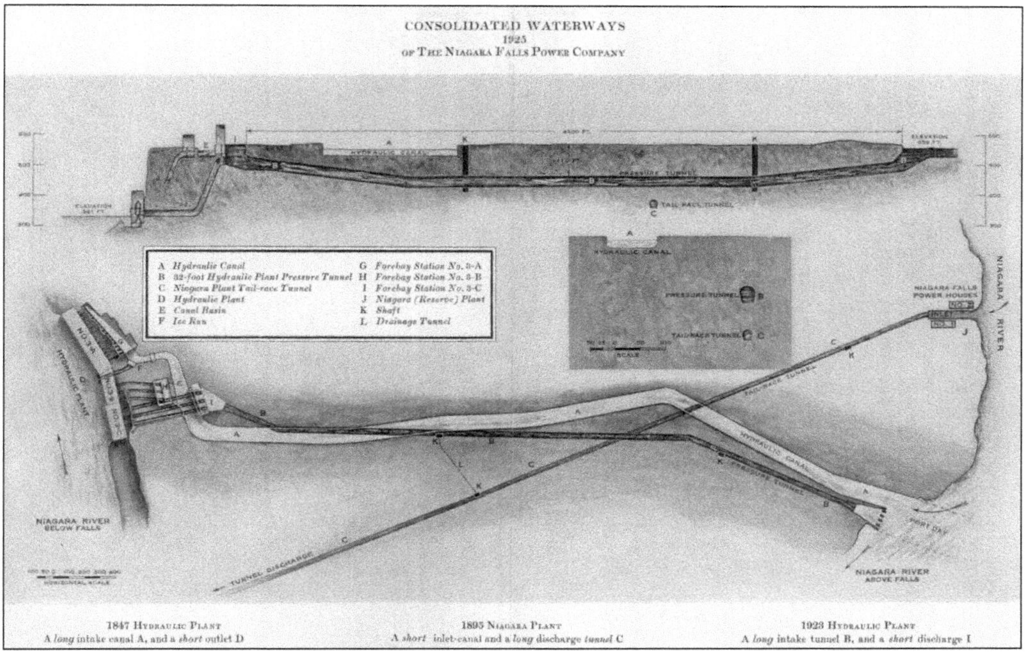

On November 16, 1896, alternating current electrical power generated by turbines capable of 100,000 horsepower was transmitted to Buffalo—more than 20 miles away—through huge underground tunnels and conduits. This map shows the Niagara Falls Power Company's canal and tunnels. (Adams.)

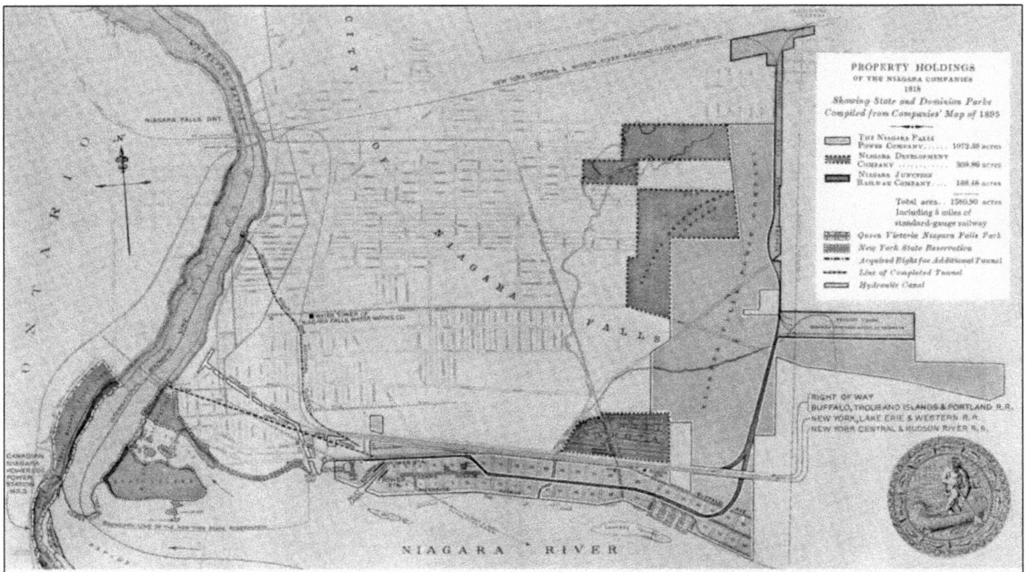

In 1899, the Cataract Company purchased the Niagara Falls Power Company. Adams wanted to entice and retain skilled workers (and their families) and to avoid labor disputes over living conditions. The Niagara Development Company was created to build housing for workers, and a 368-acre company village was built. It was called Echota, Cherokee for "a place of refuge." This 1918 map shows the holdings of the Niagara Falls Power Company. (Adams.)

Stanford White, one of America's premier architects, was hired to design the village and the houses. By 1907, Echota had 67 dwellings including small, tasteful, Colonial cottages in groupings of three to four units and single-family homes in neat rows set back uniformly from the street in a gridiron plan. All had electrical lighting, potable water, and sewerage and were designed with special attention to cross-ventilation, airflow, and natural lighting. A school, fire department, sewage disposal plant, and other community buildings were built, and the roads were paved with crushed stone and sidewalks with concrete. Echota and the ideology that inspired the building of such company towns was very well known to entrepreneurs and industrialists of the day, including William T. Love. (Both, HC.)

Two

William T. Love's Dream 1893–1912

In this May 23, 1894, photograph, John Fleming, president of the village of LaSalle, is seen "turning the first sod" for Love's project. Love is between Fleming (white beard) and a boy with a shovel. Prominent citizens, supporters, workers, horses, wagons, and other equipment are visible. A rail line alongside the canal allowed steam shovels to operate from railcars. (Photograph by J.C. Hooker, HC.)

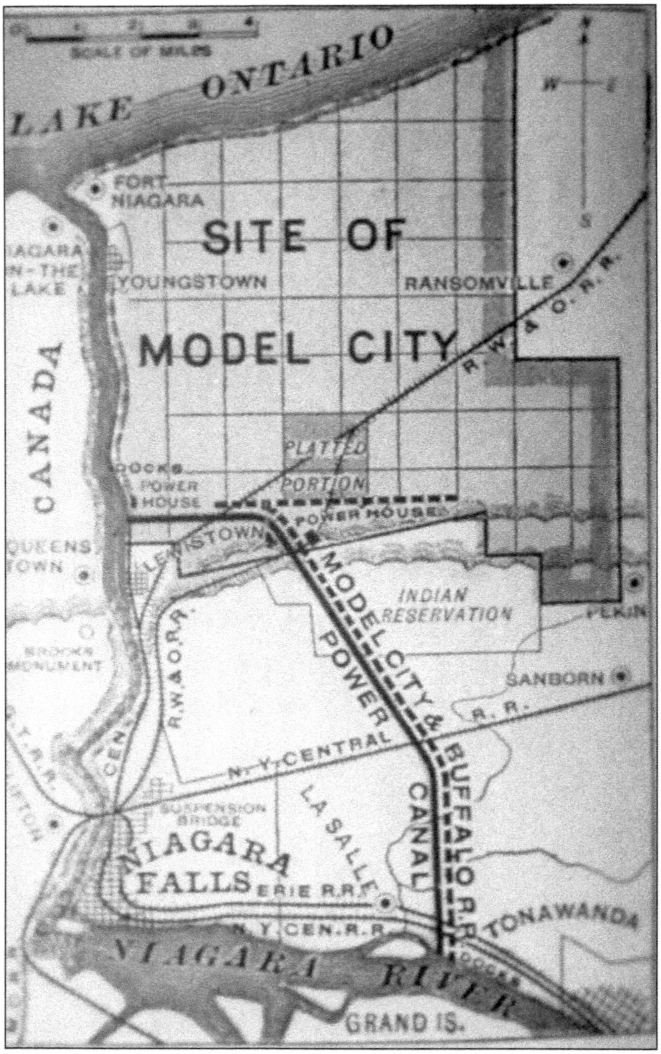

Love's model city was to be a huge city built below the Niagara escarpment, encompassing more than 30,000 acres and most of the existing townships of Lewiston, Ransomville, and Porter. Revenues from the sale of electrical power from a power canal were to fund the building and maintenance of the city. The first announcement of Love's project was in the real estate pages of the *Niagara Journal* on February 18, 1893: "A great scheme is slowly developing on the lower Niagara River . . . a plan for developing all of the village extending from Lewiston to Lake Ontario into one big city of 200,000 inhabitants." During the late 19th and early 20th century, American industrialization moved at unparalleled speed with challenges and consequences, including industrial density, crowded and unhealthy housing in cities, and poor living conditions. These conditions were unacceptable to religious and social idealists and progressive-thinking industrialists. Many reformers were guided by feelings of paternalism and the perceived need to care for workers beyond simply paying wages. Others focused upon the practicalities of retaining a skilled and contented labor force through offers of home-ownership or profit sharing. Concepts of company or model towns and of suburbia evolved from these beliefs and ideas. The ideology underlying Love's plan was the same—idealism, benevolence, paternalism, protection against labor unrest, and profit-making. (NFPL.)

Love's prospectus, titled "The Model City: Niagara Power Doubled," described the city as "the most perfect city in existence," "one of the greatest undertakings of modern times," and "the only place in the world having a great water power and deepwater navigation." Manufacturers were offered free electricity for 25 years, unlimited investment opportunities, free building sites, 60,000 affordable homes for workers, no tenements, no labor strikes, present and future shipping facilities, access to deepwater navigation, a belt railroad, cheap raw materials, vast markets, advertising advantages, fire protection, and cheap building materials. All businesses were welcome except saloons and taverns, and the sale of "demon rum" was prohibited. Prospective residents and workers were offered employment in a new method of business, a cooperative, where workers would be their own bosses and, therefore, labor problems would not occur. Residents were offered "industrial universities" that provided manual training, technical studies, and general education courses. There would also be 21,000 acres of public parks, paved streets, sidewalks, and underground sewers. (HC.)

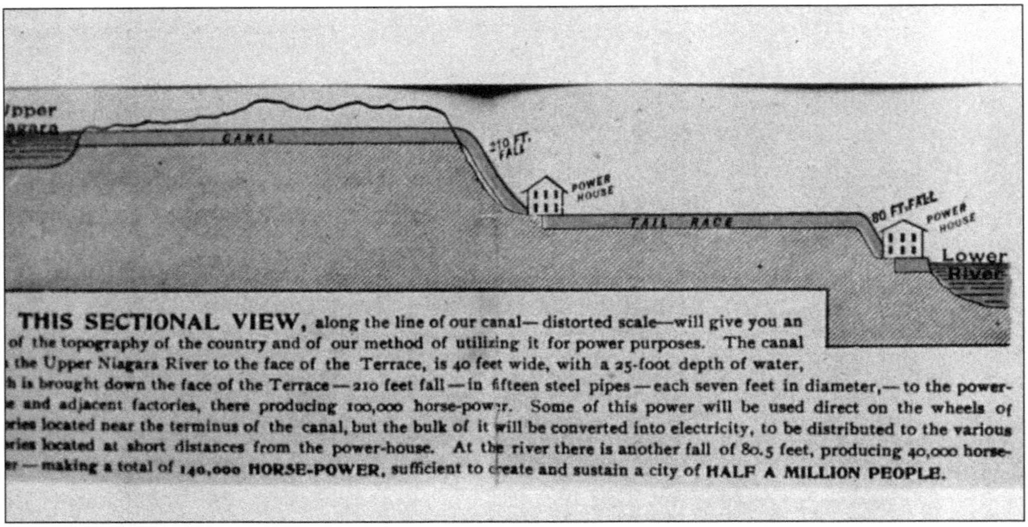

Above is a diagram of the power canal showing its path and workings. It was designed to produce direct current, not alternating current, and it was usable for shipping up to the end of navigability where the current intensified for the fall over the escarpment. The plan was to divert water from the upper Niagara River, upstream from the village of LaSalle, and direct it to the Niagara escarpment, where it would fall 210 feet to an electrical power station at the southernmost boundary of Model City. This station would provide unlimited electrical power for businesses, residents, and city services. A second section was to be dug to convey the water north and west from Model City where it would fall 80 feet to another generating facility on the lower Niagara River and provide additional power for use and sale. The map below, from Love's prospectus, shows the Welland Canal, Port Colborne, Dalhousie, and Queenston, Ontario; and Youngstown, Ransomville, Lockport, Niagara Falls, LaSalle, Grand Island, and Buffalo. (Both, NFPL.)

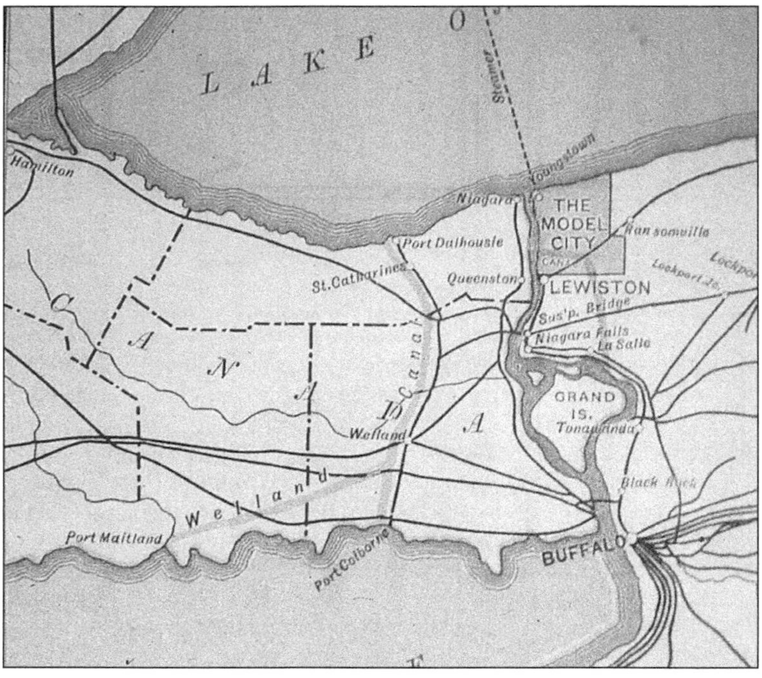

The village of LaSalle offered recreation (boating, fishing, and picnicking) and ice harvesting for both village and city residents. In 1893, a bill was introduced to the New York State Assembly by Elton Ransom to charter Love's enterprise as the Model Town Company. Love also established a company called the Niagara Power and Development Company. Ransom supported Love's project, in part, because he believed that it would turn Lewiston into a great manufacturing city. The bill gave Love's company sweeping powers, including the right to divert as much water as needed from the river; to build a town with houses, a depot, and parks; to equip and operate street railways, waterworks, gasworks, electric and steam heating plants; to operate public telephones and telegraph lines; to construct 10 miles of steam railroad tracks to connect with other railroads; and to sell sites for manufacturing. (Both, HC.)

C. A. Pooley 20 a.
Geo. Dick 50 50 a.
Reservation Line
Buchl 8 50 a.
Jno Getts
Jacob Dick 17 3 a.
E. A. Long 15 a.
Jacob Boiler 30 50 a.
F. J. Kuhlman 14 88 a.
F. Zeiger
Geo. Greenwaldt
Wm. Schultz
Aug. Millentheim
Geo. Dick 29 75 a.
Berghol
Cayuga Av.
Jno Haiser 12 50 a.
West Ave.
Elm Av.
St.
Jac. Luick
Luick Av.
Long
Jos. F.
E. Getts 6a.
E. E. Lumley 3a.
P. Heald
Wm Kidd 5 a.
Nicholas Beckwith 72 a.
Lymberner 49 a.
Eugene Smith 155 a.
Sam. Wilkinson
Hennepin Way
Pearwood
Niagara
Griffon
Road
Hy. S. Tompkins
Tompkins
Griffon St
B. S. Snodgrass
Spann & Wilcox 97 28 a.
Cutter
V A T I A
55
56
57
58
LA SALLE
Simpson St.
Sta.
Alice Latflin
Jno. J. Hopkins 50 a.
Hannah M
M.E.Ch.
La Salle Sta.
La Salle Road
River
Cayuga Creek
A. Wittenyel
H. S. Tompkins 8 a.
Lautz & Long
Cayuga Island
D. N. Long
NIAGARA SHORE TERMINAL RAIL ROAD
Steamboat Wharf

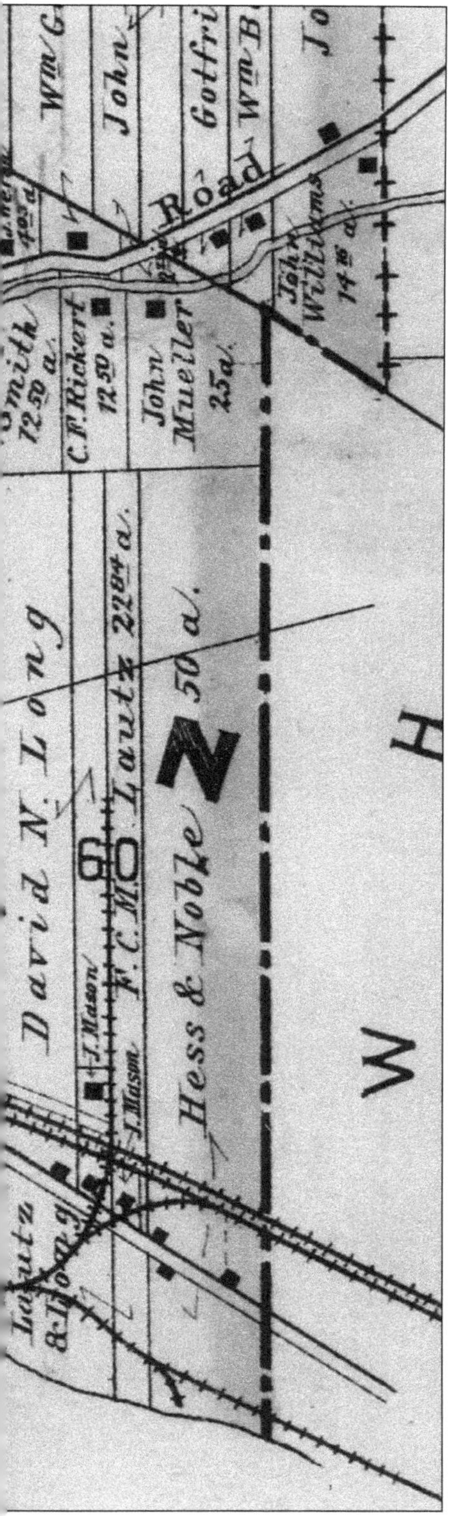

This 1893 map of LaSalle shows Cayuga Creek, Cayuga Island, and the Little Niagara River, and to the right is Lot No. 60, owned by D.N. Long. The canal was started on the west side of Long's land. The location was selected for the flat geography, natural clay, and the engineering capacities, like the ability to better divert the river at this shoreline. Love originally wanted to use Cayuga Creek for a mile or more from the mouth of the creek on the little Niagara River. However, local landowners placed too many restrictions on this use, and he decided to "take a route on which he could be independent," as reported by the *Niagara Journal*. The canal was to be 80 feet wide at the top, 40 feet wide at the bottom, 30 feet deep, and six to seven miles long.

The residents of the village of LaSalle highly anticipated the progress Love's project would bring, such as employment and prosperity beyond that afforded by the area's natural bounty of produce, fish (including sturgeon like the one pictured here), and recreation. On May 20, 1893, Gov. Roswell Flower signed the Ransom Bill. The *Niagara Journal* reported that Love had secured options on 12,000 acres and was seeking 2,000 more and that the capital stock of the company was to be $10,000,000. The incorporators included William T. Love; E.M. Love of Lewiston; Augustine Davis of Auburn Parkway, Illinois; L.L. Baily of Westboro, Pennsylvania; and William A. Wilson of Knoxville, Tennessee. (Above, NFPL; left, HC.)

This 1894 photograph shows the P.C. Flynn and Son Building where Love's Niagara Power and Development Company had its Niagara Falls offices. The three men standing in front are unidentified; however, they may be company officers William T. Love, William A. Wilson, and Augustine Davis, whose names were painted on the glass door window behind them. The second window to the right had the words "Model City" painted on it. (NFPL.)

By 1895, at least one church had been built in Model City. The *Niagara Journal* reported the name of the church as Union and the denomination of the church to be Presbyterian. (NFPL.)

By 1894, at least one factory and a number of houses had been built. These 1978 photographs show the remains of what was reported to be a nail factory and three original Model City houses. In 1895, Love's dream of a model city and power canal ended due to the loss of financial backing in the aftermath of the Panic of 1893 as well as the dramatic impact of Tesla's invention of alternating current and methods of transmitting it. By this time, Love had acquired options on more than 20,000 acres; excavated a section of canal that was 60 feet wide, 10 feet deep, and 3,200 feet long; and built a number of houses, a post office, and some factories.

The c. 1899 Model City US Post Office Building was still in operation when this photograph was taken in 1978. The Model City, New York, zip code is 14107, and the main post office is now on Model City Road.

PRICE 10 CENTS

REVOLUTION

THE FOREWORD of a PROPOSED WORLD MOVEMENT to Solve the LABOR and Other Social Problems of Our Strenuous Times. . . . .

WHICH SHALL IT BE?

INDUSTRY. PEACE. TRUTH. or FRAUD. DISCORD. VIOLENCE.

In 1906, Love tried to revive his dream and wrote a pamphlet, titled "Revolution," which proposed that the city be run by an "Agrarian Army" of worker-owners. The revival was unsuccessful, and Love reportedly moved to Ontario to explore mining opportunities. No record of Love's death or burial place can be found. The final blow to all power canal dreams also occurred in 1906, when the US Congress passed the Burton Act. This act was one of the first federal environmental laws and prohibited any further diversion of water from Niagara Falls in order to preserve them. Love's remaining properties were sold at public auction in 1910. (NFPL.)

In 1903, Elon Hooker started the Development & Funding Company in Brooklyn to manufacture bleaching power and caustic soda through the Townsend electrolytic process. In 1905, Hooker purchased the option on the Townsend cell and built a small plant in Niagara Falls. He was pursuing the American dream of entrepreneurship in what was "the golden age of invention." Niagara Falls was selected because it was within 60 miles of the largest salt mine in the western hemisphere, it was on the shore of the Niagara River with access to the greatest water-powered electrical source in the world, and it was adjacent to the Great Lakes transportation system. The first plant was on a 6.75-acre pear orchard with a farmhouse and barn at the corner of Buffalo Avenue and Union Street. By 1906, the plant was producing 11 tons of bleaching power and five tons of caustic soda a day. In 1909, the company name was changed to the Hooker Electrochemical Company (Occidental Petroleum acquired it in 1968). (University of Rochester.)

The population of the village of LaSalle grew with the expansion of industry in the city of Niagara Falls. It became a suburb and was annexed in 1927. The village, Cayuga Island, Cayuga Creek, and the Little Niagara River continued to be popular recreational destinations. This 1914 photograph shows boathouses on the Little Niagara. (NFPL.)

Three

Sowing the Seeds of a Nightmare 1913–1952

This 1938 aerial photograph shows Love's canal excavation, with the piles of excavated soil. The photograph also shows that the canal was connected to the Niagara River. The Little Niagara River and Cayuga Island are in the lower right, and roads and railroad lines are running across the bottom, parallel to the Niagara River shoreline. In 1941, Hooker initiated a feasibility study to determine whether Love's ill-fated power canal was suitable for the disposal of chemical waste. The site was determined to be quite suitable due to the fact that the canal had been dug in an area of virtually solid clay. The naturally impermeable nature of clay had been recognized as useful in entombing waste and was the state of the art of industrial waste disposal. Hooker's manufacturing processes produced wastes. These wastes were not simply trash and garbage—these were industrial and chemical wastes. (WPC.)

The late-1930s photograph above shows boys swimming in the abandoned canal. Old-timers recall that the canal section was not landlocked, as some would assert in the 1970s, and was actually connected to the Niagara River through excavation or small creeks. It was possible to navigate small boats or canoes from the old canal to the Little Niagara River, Cayuga, Black, and Bergholtz Creeks and back again. In April 1914, the beginning of World War I, the Hooker Electrochemical Company produced two products: bleach and caustic soda. By the time World War I ended in 1918, Hooker was producing 15 additional chlorine-based products, and in 1915, it built the first US monochlorobenzol plant. Monochlorobenzol is a coal tar intermediate used in the making of muriatic and picric acids. Benzol, phenol, and toluol are also extracted from coal tars and chlorinated to form intermediates. These chemicals were used in the production of explosives. In 1918, the Hooker plant was the largest in the world. Below is a 1915 photograph of boaters on Cayuga Creek. The abandoned section of Love's canal had lain fallow for 20 years. Over time, it had filled with water from the many small creeks, streams, and swales that crisscrossed the area, and it came to be used for a new purpose—recreation. It was a swimming hole and a pleasant place for picnicking, boating, and fishing in summer and for skating and ice fishing in winter. (Above, LCARA; below, NFPL.)

This 1942 US War Department map shows the canal excavation at center, the piles of excavated soils, and the work extending to the Niagara River. In 1942, Hooker entered into an agreement with the Niagara Power and Development Company (Love's former company) to use the 16-acre canal as a dump site for industrial waste. Also in 1942, the Lake Ontario Ordnance Works, a US Government War Department project that made explosives such as TNT, opened a plant in Model City. Hooker sold various materials to the works over the years. From 1942 to 1952, Love's former canal was put to a new and ominous use—the disposal of more than 21,000 tons of chemical wastes, including chlorinated hydrocarbon residues, sludge, fly ash, and solid wastes from the production of solvents, dyes, defoliants, pesticides, explosives, and other products. Hooker's wastes were disposed of in used metal drums, barrels, and fiber containers. Many of these containers were themselves garbage. Hooker also pumped chemical waste from tanker trucks in liquid form or as sludge. (WPC.)

Visible in this 1942 aerial photograph is the water-filled canal running north (right) to south (left) and connecting with the river (visible as the black vertical line). To the right of this line is the 102nd Street dump site. A small creek, running from Bergholtz Creek at the north, clearly crosses Colvin Boulevard and the northern section of the canal. The eastern section of Cayuga Island is visible at the south as well as the Little Niagara River, Buffalo Avenue, Frontier Avenue, and rail tracks. Hooker excavated and backfilled various areas of the canal, and residents recalled the smell of chemicals, the sight of fly ash in the air, the death of their lawns and trees, and the appearance of burned paint on their houses. Some residents even remember using their garden hoses to wash off workers who were splashed by chemicals. (WPC.)

By 1952, the canal was full, and Hooker covered it with soil and grass grew. While it appeared to be a harmless meadow, the seeds of an eventual environmental and public health nightmare had been sewn. In 1943, Hooker began dumping south of Love Canal at its 102nd Street site. This site was crossed by creeks and swales and is only a quarter of a mile upstream from the city's drinking water intakes.

By 1971, Hooker had dumped more than 23,000 tons of chemical waste here. By 1976, toxic chemicals (including Hooker's Mirex) had been detected in Lake Ontario fish and were traced back to dump sites on the Niagara River—including this one—and the resulting investigations led to Love Canal. The "No Trespassing" sign and creek were photographed in June 1978.

Four

American Dream 1953–1975

The berm of the old canal is visible behind the children in this late-1940s to early-1950s photograph. In 1953, the 16-acre site was sold to the Niagara Falls School Board for $1. Hooker demanded a quit claim deed that reads in pertinent part as follows: "The premises . . . have been filled . . . with waste products resulting from the manufacturing of chemicals. . . . The grantee assumes all risk and liability incident to the use thereof. . . . No claim, suit, action or demand of any nature whatsoever shall ever be made by the grantee, its successors or assigns, against the grantor . . . for injury to a person or persons, including death resulting there from, or loss of or damage to property caused by . . . industrial wastes. It is further agreed as a condition hereof that each subsequent conveyance of the aforesaid lands shall be made subject to the foregoing provisions and conditions." (LCARA.)

This 1951 aerial photograph of the Love Canal neighborhood shows the ongoing development. In 1952, the Niagara Falls City School Board determined that population growth necessitated the building of another public elementary school in the area and contacted Hooker regarding the site. Hooker executives warned the board about the chemical wastes buried in the canal and the unsuitability of the area for building and basement excavation. (WPC.)

Hooker's warnings, conditions, and provisions regarding the site were part of the school board's records, but this information was not known by the general public and not conveyed to those who purchased land adjacent to the dump site from private developers. The board agreed that the school would not have a basement or an inground swimming pool, that Ninety-eighth Street would not be extended across the site, and that the rest of the site would be used for a public park and not for home building. The new, modest, single-family homes, park, and elementary school were all part of the postwar American dream of suburbia and great enticements for young families. Many of those moving to the neighborhood were blue-collar workers employed by the chemical industries of Niagara Falls, including Hooker Chemical.

Niagara Falls zoned the area as general residential, and the news that a new elementary school and public park were going to be built brought a wave of land purchases and home building. As evident in this 1978 aerial photograph, an entire neighborhood was constructed around the Ninety-ninth Street School. The new neighborhood offered affordable single-family homes with garages and backyards. Seeking their piece of the American Dream, many young families with school-age children were attracted to the neighborhood. In 1955, the Ninety-ninth Street Elementary School opened its doors to 400 children, most of whom walked to school and went home for lunch. (WPC.)

The board initially planned to have the school constructed directly in the center of the former dump site but was forced to move the plans 75 feet to the east when a construction crew hit chemical waste during excavation for the concrete slab. In addition, it was decided that the school would have a tile drain system installed around the slab to collect and divert groundwater away from the school. This drain system was connected to the storm sewers and allowed contaminated groundwater and liquid chemical waste (leachate) to flow untreated and unabated into the Niagara River until 1978. In August 1953 and January 1954, the board authorized the removal of thousands of yards of topsoil from the surface of the site for grading at the Ninety-third Street Elementary School. These actions would later contribute to the deterioration of the clay cap and contaminate the Ninety-third Street School grounds. (WPC.)

This 1956 aerial photograph shows the Ninety-ninth Street School in the center, Ninety-seventh to Ninety-third Streets to the west (top), and Ninety-ninth to 103rd Streets to the east (bottom). The Griffon Manor public housing development was built to the west of the site, and in 1971, the LaSalle Housing Development and senior center were built. This housing was very popular, for it was new and offered the benefits of suburbia, including garden apartments, parks, and the Ninety-third Street Elementary School. (WPC.)

Children play in a backyard sprinkler with the canal landfill visible in the background around the 1960s. (SUNY.)

Five

The Nightmare Unfolds 1976–May 1978

In 1974, Karen Schroeder, shown here with a jar of black chemical sludge in her hand and the school in the background, looked out the window of her Ninety-seventh Street home and saw her fiberglass inground swimming pool two feet out of the ground floating on top of a chemical-smelling liquid. What began as a property loss for her was the beginning of a nightmare of disastrous proportions for hundreds of her Love Canal neighbors. (SUNY.)

In 1976, David Pollak, a *Niagara Gazette* reporter, investigated a homeowner's repeated complaints about having to replace his sump pump after it clogged on a thick, black, acrid-smelling substance. Pollak scooped up a jar of the sludge to have it analyzed, and a signature Hooker compound called C-56 was detected. In 1977, the city hired a consultant to start hydrogeological testing, and in early 1978, the New York State Department of Environmental Conservation (DEC) began intensive testing. In April 1978, the New York State Department of Health (DOH) called the Love Canal "an extremely serious threat to health and welfare" and ordered the Niagara County Health Department to begin health testing and restrict access to the site. This photograph shows the four-foot-high snow fence and danger sign that were erected.

The Ninety-ninth Street School would remain open for the remainder of the school year, but the playground equipment was dismantled, and the playing fields and baseball diamonds were enclosed by snow fence. Parents began to wonder whether the school would reopen in September 1978 and whether they should send their children there. These parents included Lois and Harry Gibbs, who lived on 101st Street with their school-age son and younger daughter. In April 1978, ninety-seven families, with 230 adults and 134 children, occupied the homes immediately adjacent to the dump site.

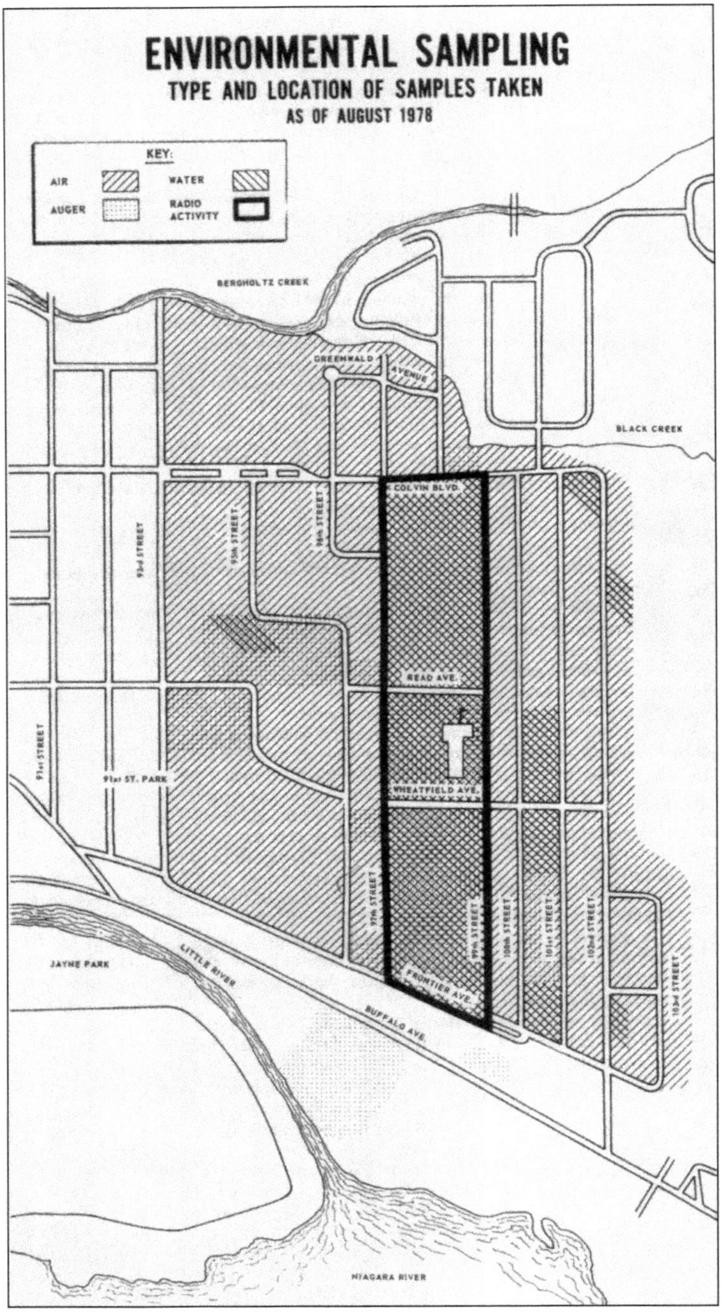

The DEC and DOH tested air, soil, and groundwater for chemicals and radioactivity. Tests in the basements of 11 homes adjacent to the dump site revealed the presence of a variety of compounds. The problem was not limited to a few homes—a potential public health hazard existed. Concern about the potential health hazards led to prioritizing basement air sampling of all homes adjacent to the site before testing soils, ground, and surface water. Many of these "Ring 1" homes had finished basements, which were used as bedrooms, and officials focused upon minimizing residents' exposure through their inhalation of toxic vapors. (DOH.)

TABLE I

POTENTIAL HEALTH EFFECTS OF CHEMICAL COMPOUNDS IDENTIFIED AT LOVE CANAL

COMPOUND	ACUTE EFFECTS	CHRONIC EFFECTS
benzaldehydes	allergen	
benzene	narcosis skin irritant	acute leukemia aplastic anemia pancytopenia chronic lymphatic leukemia lymphomas (probable)
benzoic acid	skin irritant	
carbon tetrachloride	narcosis hepatitis renal damage	liver tumors (possible)
chloroform	central nervous narcosis skin irritant respiratory irritant gastrointestinal symptoms	
dibromoethane	skin irritant	
dioxin	chloracne	nervous system disorders psychologic abnormalities cancer, spontaneous abortions, liver dysfunction (indicated in animal studies)
lindane	convulsions high white cell counts	
methylene chloride	anesthesia (increased carboxy hemoglobin)	respiratory distress death
trichloroethylene	central nervous depression skin irritant liver damage	paralysis of fingers respiratory and cardiac arrest visual defects deafness
toluene	narcosis (more powerful than benzene)	anemia (possible) leukopenia (possible)

On May 15, 1978, the DOH announced that air samples taken inside homes adjacent to the site detected 80 chemical compounds and dangerously high levels of toxic vapors in the basements of homes in the first ring (Ninety-seventh and Ninety-ninth Streets). Over 200 distinct chemicals would be identified, 11 of which were known or suspected animal carcinogens and one of which was the known human carcinogen benzene. Also on May 15, New York State officials meet with concerned residents in the auditorium of the Ninety-ninth Street School to provide information on the state's plan to contain and cleanup the area. On May 19, a state toxicologist met with residents at the school and attempted to explain the hazards of exposure to toxic chemicals. Residents were alarmed by the complex health information and by officials and scientists who did not have answers to their questions and seemed to be focused on cleaning-up and containing the site. (DOH.)

Testing of air, water, and soil for toxins continued. Sumps were tested, including this one in the basement of a Ninety-ninth Street home. Neighbors who had previously compared family illnesses and property damage over their backyard fences began discussing the testing going on in their neighborhood and feared the extent of their family's exposure. After reading disturbing articles in the *Niagara Gazette*, Lois Gibbs started a petition drive to close the Ninety-ninth Street School.

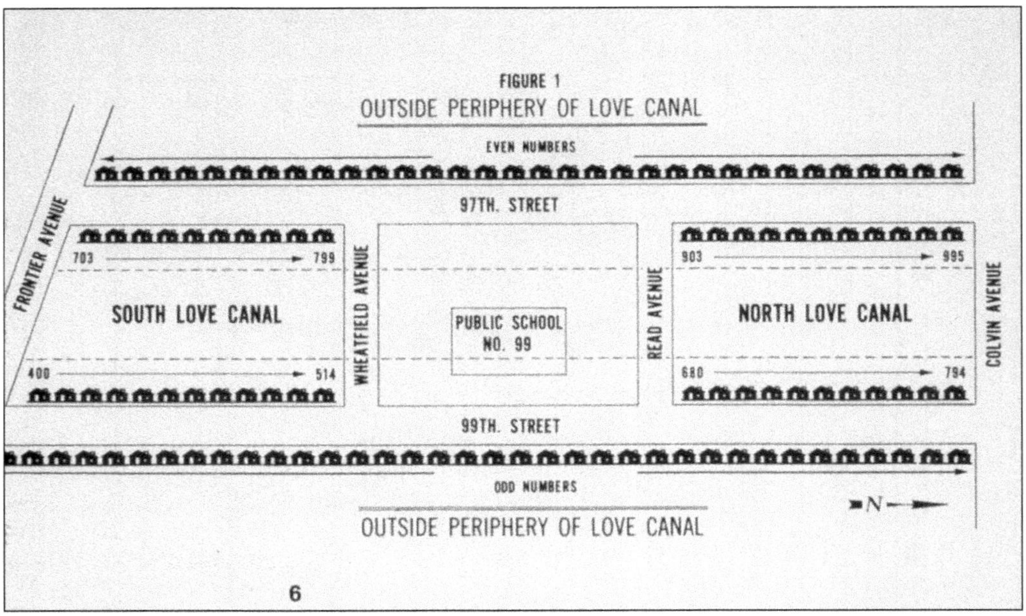

As shown on this map, the state initially defined the boundaries of the Love Canal area as a rectangle enclosed by existing streets. This definition would be central to one of the Love Canal controversies: the boundary lines for the relocation of residents. Initially, the state defined Ring 1 (homes bordering the canal); then, Ring 2 (homes across from Ring 1 homes); and finally, Ring 3 (the outer streets and declaration area). (DOH.)

Six

Awakening, Awareness, and Response June 1978–October 1980

In June 1978, the Love Canal neighborhood was established and thriving. Well-maintained, modest one-family homes on tree-lined suburban streets had patios and swimming pools in their backyards, children's toys and campers in the driveway, and US flags flying. Many were starter homes for young families, some were built onto by growing families contented to stay, and some were owned by retired couples attracted to the pleasant neighborhood.

Ninety-ninth Street looks serene in these summer 1978 photographs. Children are walking on the sidewalk, the trees are full, lawns are mown, American flags are flying, and flowers are in bloom. Many Love Canal neighborhood residents were veterans, with firm commitment to country and faith in government and the American way. (Below, WPC.)

By the summer of 1978, the southern section of the canal (especially behind Ninety-seventh Street homes) was eroded and had considerable visible contamination. This 1978 aerial photograph shows the southern section of the neighborhood bordering on Frontier Avenue. Homes on Ninety-seventh Street with yards bordering on the dump site are at top of the photograph, and homes on Ninety-ninth Street with yards bordering on dump site are at bottom of the photograph. Every home was extended to the limit of its property line and many had swimming pools and patios. The barren sections on the surface of the dump site are where the topsoil had been removed or eroded and chemical waste (either in rusted, collapsed barrels or in sludge form) had come to the surface. The bottom aerial photograph shows the southern section of the site farther northward. (Both, WPC.)

These 1978 aerial photographs show the condition of the southern end of the site moving north towards Wheatfield Avenue. Wheatfield Avenue is seen to the right as well as one of the school's baseball diamonds. (Both, WPC.)

This 1978 aerial photograph shows the center section of the site bordered by Wheatfield Avenue. One of the school baseball diamonds, the playground, and the southern half of the school are also visible. (WPC.)

This 1978 aerial photograph shows the center section of the site moving northward. The Ninety-ninth Street School, another of the baseball diamonds (with a person walking across it), and the school's parking lot are visible. At the top of the photograph are houses on the opposite side of Ninety-seventh Street. (WPC.)

This 1978 aerial photograph shows the center section of the canal site bordering on Read Avenue with the northern section of the site to the right of Read Avenue. The parking lot and baseball diamond of the Ninety-ninth Street School are visible, and so are areas of scarce vegetation and what appears to be part of a mini-bike track behind houses on Ninety-ninth Street. (WPC.)

This 1978 aerial photograph show the northern section of the site that borders on Colvin Boulevard. Ninety-seventh Street houses with yards bordering on the dump site are visible at top of the photograph, and houses on Ninety-ninth Street with yards bordering on dump site are at bottom of the photograph. (WPC.)

What lay behind these Love Canal dream homes, just beyond their property lines, was a pockmarked scene of collapsed rusty barrels emerging through the surface; pools of black oily sludge; and an omnipresent chemical stench. At night, residents reported sometimes seeing an eerie green ground fog over sections of the canal's southern send.

Rusty barrelheads and rims, holes where barrels had collapsed, and groundwater mixed with leachate were present on the surface of the former dump site, directly behind Ninety-seventh Street homes, in the summer of 1978. The site was not fenced off, and children still played there.

Holes, rusty barrel rims, and collapsed barrels filled with chemical leachate, as well as solid waste and burned vegetation, were present on the southern end of the canal in the summer of 1978. Between 1942 and 1953, Hooker disposed of an estimated 200 tons of trichlorophenol (TCP); 400 tons of acid chlorides (acetyl, caprylyl, butyrl, and nitro benzoyls); 400 tons of metal chlorides; and 500 tons of thionyl chloride and sulfur/chlorine compounds in liquid and solid form in drums. Dioxin, one of the most toxic man-made compounds, is a by-product of the manufacture of TCP, and it was found in storm sewers, sediment, leachate, and soil samples from the canal surface and backyards.

Pools of standing water, piles of debris, and collapsed barrelheads were visible directly behind homes on Ninety-seventh Street in the summer of 1978. Between 1942 and 1953, Hooker disposed of an estimated 700 tons of liquid disulfides and chlorotoluenes; 800 tons of benzoyl chlorides and benzotrichlorides; and 1,000 tons of chlorinated waxes, oils, naphthenes, and aniline in liquid and solid form in drums.

A rusted-out barrelhead with a jagged rim and what appeared to be solid waste and the remnants of a fiber barrel were on the southern section of the canal in the summer of 1978. Between 1942 and 1953, Hooker disposed of an estimated 2,400 tons of dodecyl mercaptans, chlorides, and organic sulfur compounds; 2,400 tons of benzylchlorides; and 2,000 tons of sodium sulfide/sulfhydrates.

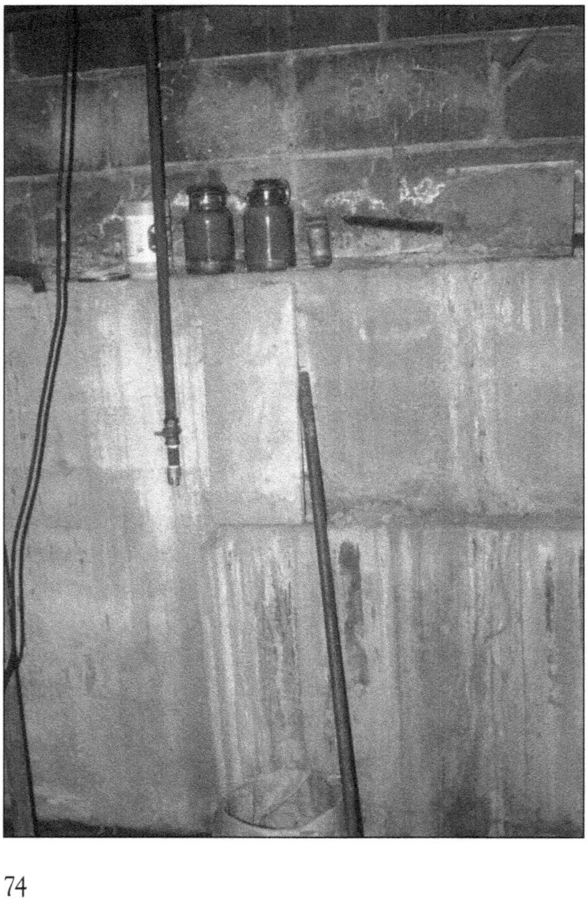

A deteriorated, collapsed fiber barrel was visible on the surface of the canal in the summer of 1978. Between 1942 and 1953, Hooker disposed of an estimated 6,900 tons of hexachloro-cyclohexane (Lindane) and an estimated 2,000 tons of clorobenzenes in liquid and solid form in nonmetallic containers and drums.

Chemical waste leaking from the dump began breaching foundations and permeating hollow cinder block walls in many basements. A basement wall of a house on the southern section of Ninety-ninth Street had streaks of leachate running down the wall. Many homeowners had tended backyard gardens and canned their own vegetables, and two jars of canned tomatoes were stored on a ledge in this home.

Many Love Canal homes had bedrooms or family rooms in their basement. Here, a finished basement in a south Ninety-ninth Street home had chemical leachate coming through the walls; a fishing rod lies abandoned on the floor.

The basement of a home on south Ninety-ninth Street had leachate come up through the basement storm drain. At the top of this photograph, the sump pump can be seen. It was filled with leachate. All Love Canal homes had basement sump pumps, drains, and septic systems, because municipal sewer lines had not yet been extended to the suburb. In addition, some homes used well water.

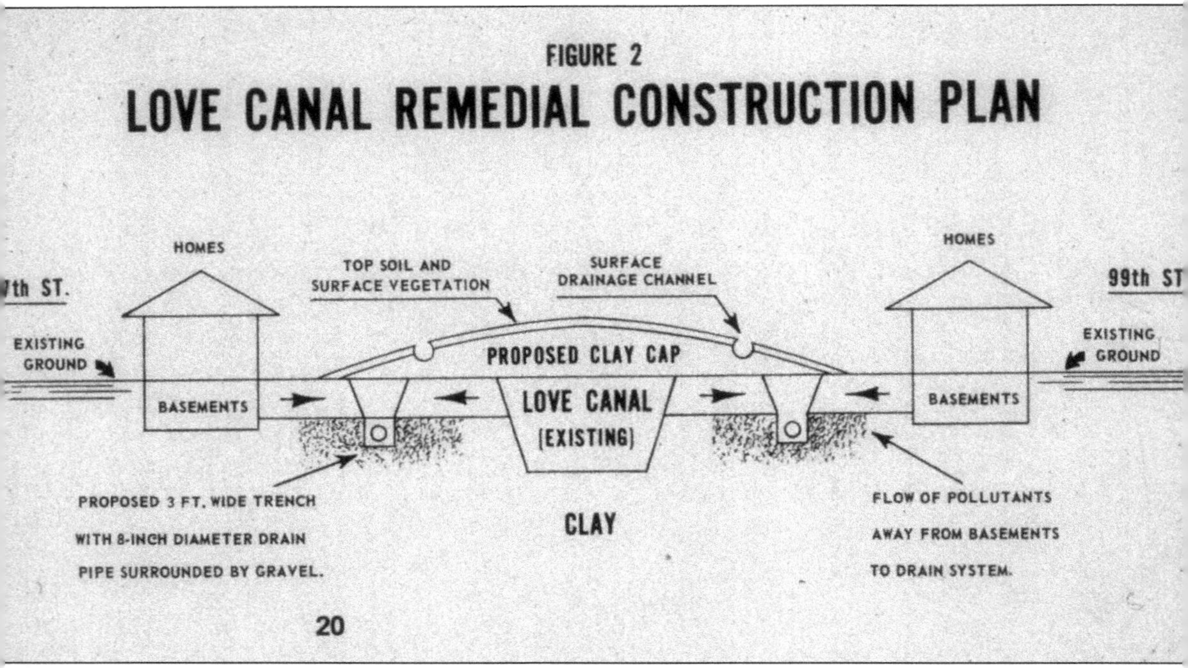

On June 13, 1978, state officials met with residents to explain the remedial construction and containment plans. As this diagram indicates, the plan was to contain the waste and prevent any farther migration of it by trenching north and south along the presumed rectangular old canal boundaries, laying perforated clay drain tile to collect leachate and groundwater from the site, installing an impermeable clay cap over the top of the area, and installing monitoring wells. The trenches were not to be dug laterally (west and east) across the dump site or down to the bedrock. The remedial plan would become central to major Love Canal controversies involving relocation lines and an alternative theory of chemical migration and exposure known as the swale theory. (DOH.)

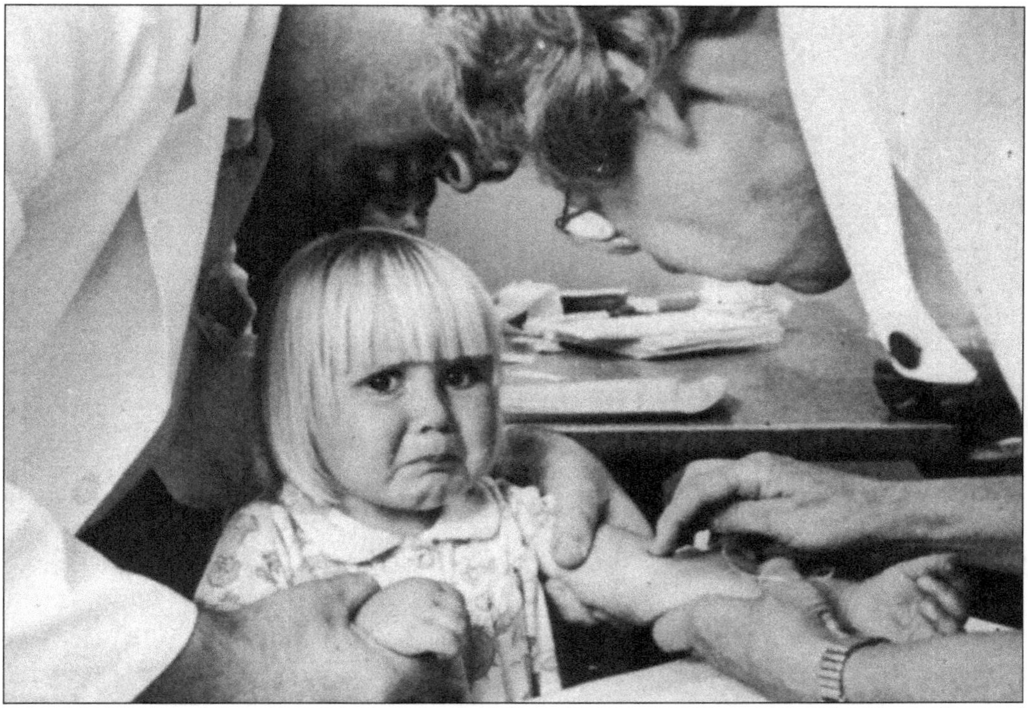

On June 15, 1978, the *Niagara Gazette* reported that a high rate of birth defects had been detected in the neighborhood. On June 19, the DOH began a house-to-house health survey of the 97 families living adjacent to the dump site and collected blood samples from residents, including children. (DOH.)

On July 7, 1978, the DOH released the results of air sample analysis collected from basements showing a high level of toluene, chlorotoluene, and chloroform. Local and state officials met with confused and frightened residents in the school auditorium to explain the data. This photograph shows Niagara Falls mayor Michael O'Laughlin speaking to residents in the school auditorium while holding a child in his arms.

On August 2, 1978, New York State health commissioner Robert Whelan declared a health emergency at Love Canal and advised pregnant women and children to leave the area. No provisions were announced for an evacuation, and residents were not told how long they would have to leave for or what would be done to resolve the situation. Gov. Hugh Carey appointed the Love Canal Interagency Task Force (LCITF) and told residents that the state would pay for temporary housing until their children were two. The *New York Times* published a front-page article, "Upstate Waste Dump Troubles Neighbors," thereby placing Love Canal in the national and international spotlight. On August 4, the LCITF opened an office in the Ninety-ninth Street School, and homeowners, terrified and outraged by the limited temporary relocation, formed the Love Canal Homeowners Association (LCHA) and elected Lois Gibbs as president.

On August 7, President Carter declared a federal emergency and authorized actions to be taken to "save lives, protect property, and avert or lessen the threat of disaster." On August 9, Governor Carey met with Lois Gibbs and announced that the state would evacuate all 239 families living on both sides of Ninety-seventh and Ninety-ninth Streets (Rings 1 and 2) and buy their homes. This was the first permanent relocation. (DOH.)

On August 22, 1978, the state began replacing snow fence with an eight-foot cyclone fence around Rings 1 and 2 homes. The fencing cost $58,000 (the value of approximately two homes at the time), and it became a focal point for controversy and protests involving health damages, relocation boundaries, lost property values, and the migration of toxic chemicals.

Many residents watched as the fencing encircled Rings 1 and 2, becoming increasingly alarmed and angry. On August 30, 1978, the school board closed the Ninety-ninth Street School pending the outcome of testing for chemical contamination.

Fencing was installed behind a Ninety-ninth Street home that had been in the process of being expanded for perhaps a family room or an additional bedroom. This fencing was directly in front of "fenced-out" Ring 3 homes on 100th Street that were not part of the state's emergency declaration area.

Residents on the outside of the fence found it difficult to resume their normal daily lives. Ring 1 homes on Ninety-seventh Street at Frontier Avenue can be seen through the perimeter fence in September 1978.

A boy walks along Frontier Avenue running his hand along the Rings 1 and 2 perimeter fence that enclosed the Ninety-ninth Street School and playground. Children were transferred to the Ninety-third Street Elementary School when the Ninety-ninth Street School was closed in September 1978.

A double arrow sign is indicating no through traffic across Wheatfield Avenue in September 1978. A school sign depicting children carrying books is visible through the fence, as are playground equipment and the Griffon Manor public housing project in the distance. Griffon Manor is shown on the other side of the fence that separated the Ninety-seventh Street Ring 1 and 2 homes. In September 1978, some of those who rented homes or lived at Griffon Manor formed the Concerned Love Canal Renters Association (CLCRA).

Many 101st Street residents living across the street from the fence became outraged by the unclear, yet significant enough, health threats to necessitate the fencing as well as the decimation of their property values due to their proximity to the apparent dangers. They placed protest signs on their property. Through their protest signs, residents spoke to their elected officials, government agencies, the news media, and anyone else who drove by, asking whether the fence would enclose the migrating toxic chemicals or keep their families safe.

On September 7, 1978, the DOH released a report titled "Love Canal Public Health Time Bomb" describing the area as "an environmental nightmare" capable of causing "profound and devastating effects" on health and the environment and posing a situation of "great and imminent peril." More than 82 hazardous chemical compounds were identified, including known or suspected carcinogens like benzene. (Right, DOH.)

Residents expanded their protests, staged many rallies and marches, and sought the assistance of environmental activists and the news media. (Above, DOH.)

The LCHA raised concerns about the safety of residents and workers during the remedial construction, and an evacuation plan that included an alarm siren, evacuation meeting places, and buses idling at those sites during work hours was established. In case the alarm went off, residents were instructed to have a suitcase packed with essentials next to their door to bring with them to the evacuation sites. Residents looking across the street from their homes saw these signs and were upset because the danger was so close and yet they were beyond the official boundary for relocation and, therefore, not endangered. On September 9, 1978, the DOH detected radioactivity at the Ninety-third Street School, and on September 12, the DOH extended the health testing area from Ninety-third to 103rd Streets.

Evacuation buses lined up every workday at evacuation pick-up areas outside of Rings 1 and 2. They remained in place, with engines idling, until the work ended each day at a cost of $30,000 per week. The smell of diesel fuel permeated the neighborhood. Many worried whether their elderly or infirm neighbors would make it out of their homes and to the evacuation sites in time.

The evacuation plan was extended to Ninety-sixth Street. Residents continuously listened for alarm sirens, and many could not bear to see the fence, buses, and evacuation signs and either closed their drapes or left home during the day. Many worried about explosions or spills occurring at night. Residents experienced a range of feelings and emotions including anxiety, depression, betrayal, stress, abandonment, shame, despair, and anger. These November 1978 photographs show an evacuation bus (left) with a Griffon Manor apartment building (right) and Ninety-seventh Street homes in the distance.

These November 1978 photographs show the Griffon Manor apartment buildings and evacuation signs. Residents, be they homeowners or renters, all felt profound guilt and responsibility for having unwittingly placed their families in jeopardy because of where they choose to live. Homeowners despaired over the loss of their homes and what those homes represented: their primary asset and life savings and their independence, self-sufficiency, and prosperity. Many grew increasingly angry that they were kept in the "waking nightmare" of daily fear for their lives due to official definitions of the boundaries of Love Canal.

On September 19, 1978, Lois Gibbs presented the DOH with a map of the neighborhood on which known creeks, irrigation ditches, and old stream beds or swales had been drawn as well as marking the location of homes with residents suffering from specific illnesses in different colored inks. The LCHA had collected detailed health data through telephone interviews with most residents. The LCHA believed its data supported an alternative to the state's concentric rectangular theory of chemical migration from the site. The LCHA also gave the DOH a list of 57 outer-ring families with significant health issues living near swales. The DOH rejected the LCHA's theory, calling it "unscientific" and "useless housewife data." (SUNY.)

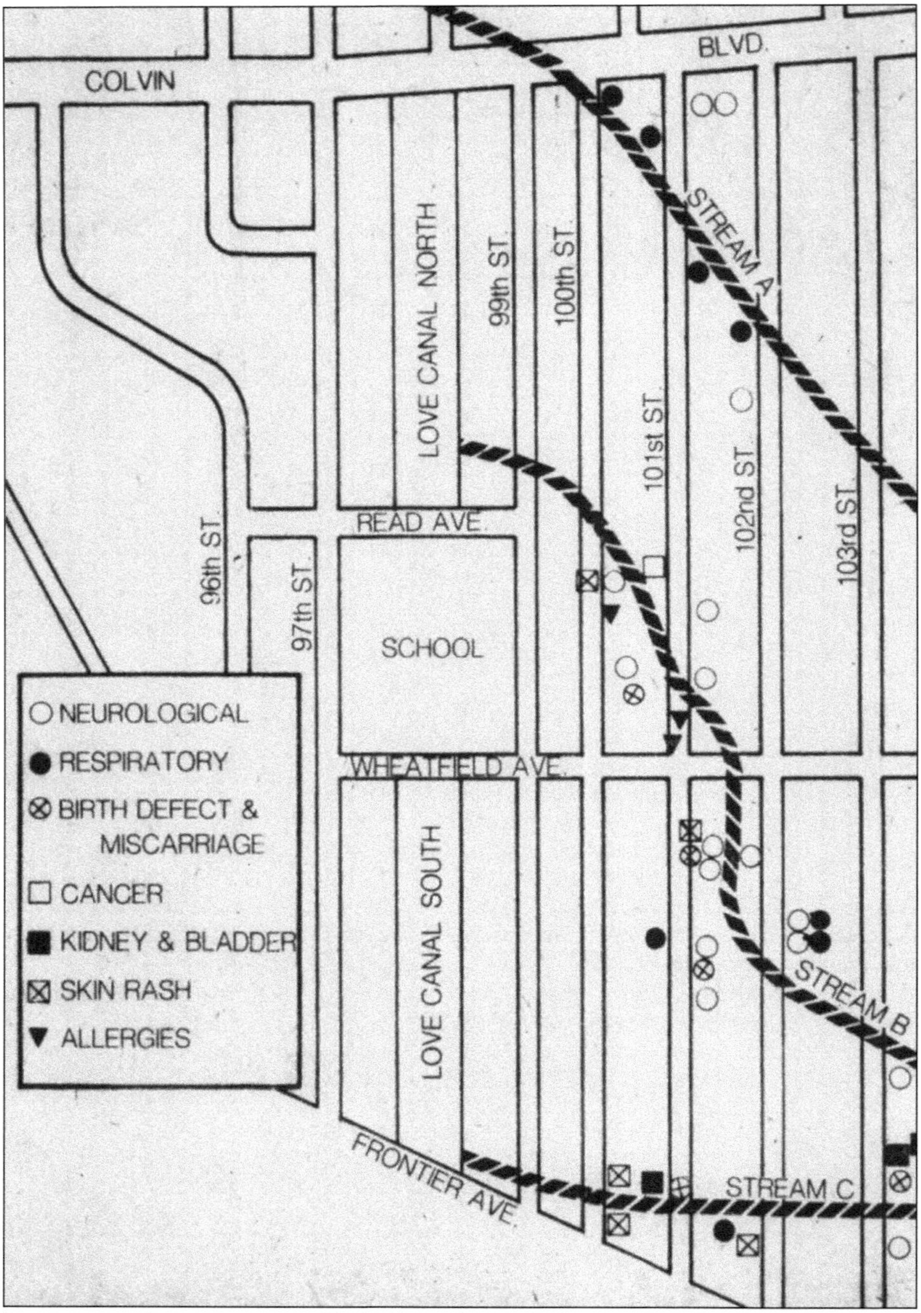

Local and regional newspapers published articles with diagrams of the LCHA's swale theory, and after additional DOH study, the swale theory of chemical migration became the accepted explanation for why some outer-ring residents had as many significant health damages as the inner-ring residents. The LCHA also determined that the miscarriage rate before living in the Love Canal area was 8.5 percent, but it was 16.5 percent after moving to the area and a shocking 25 percent in the swale areas. The news coverage of the swale theory established the LCHA as a legitimate newsmaker and a reliable, independent information source. (SUNY.)

On October 10, 1978, the remedial construction project began. The goal was containment of the hazardous waste—not excavation and disposal elsewhere. The first step was replacing the temporary snow fence the city had erected to keep people out of the dangerous areas with five-foot fencing and warning signs. The second step was erecting a temporary chemical waste treatment facility in the southern end of the site.

An EPA mobile unit and trenching machinery were moved onto what had been a Ninety-ninth Street School playground (the Ninety-ninth Street School is in the background). Remediation work began behind Ninety-seventh Street homes on the southern end of the site, where surface conditions were the worst and contamination most evident.

Work began on the southern section of the canal with large backhoes beginning to dig the trench for the remedial drain system. Boarded up Ninety-seventh Street homes are visible in the background in October 1978.

The roof of Griffon Manor is visible in the far left background and Ninety-seventh Street homes and an EPA testing unit are visible in the foreground of this photograph of south-end remediation activities in October 1978.

The backs of boarded-up and fenced-off south-end Ring 1 homes are seen in this October 1978 photograph as well as abandoned playground equipment and remediation equipment.

On October 12, 1978, Lois Gibbs telephoned the DOH and asked about 57 families with serious health problems who had requested relocation from Ring 3. She was told that they would not be relocated because they were not within the official relocation area. On October 16, the LCHA organized a protest march and picketed Niagara Falls City Hall demanding further relocations. This was the first of many protest marches and demonstrations the residents of Love Canal would participate in over the course of the next two years. The LCHA began to plan media events that paralleled national and international events. For example, there were the flotillas of Cuban and Vietnamese refugees that inspired to residents to call themselves the "Love Canal boat people."

The remedial work continued with all deliberate speed through the fall of 1978 in an attempt to complete the trenching, drainage tiling, and clay capping of the southern end before the spring thaw of 1979.

Remedial work continued from dawn to dusk, rain, snow, or shine, throughout the fall and winter of 1978–1979.

The containment plan called for capping the entire site with thousands of tons of clay, and dump trucks began bringing it onto the site in a continuous caravan, day after day, for weeks.

The clay capping was gradually built up on the southern section of the canal. Eventually, it would be extended over the remedial containment trenching and the entire dump site and then tapered and seeded so as to appear as a slight hill.

Eventually, the clay capping on the southern section of the site was as high as the eaves of the abandoned homes on Ninety-seventh and Ninety-ninth Streets.

In November 1978, testing revealed abnormal liver functions in residents living beyond Rings 1 and 2. In December 1978, residents learned that Hooker Chemical had dumped as much as 200 tons of dioxin in the site. Dioxin is produced in the making of the herbicide trichlorophenol and is one of the most toxic substances ever created. In 1976, a small airborne release in Seveso, Italy, resulted in mass evacuation and the scraping off of contaminated topsoil over hundreds of acres. Residents, fearing that clay contaminated with dioxin was being spread throughout the neighborhood, blocked trucks from entering and leaving the site. Numerous protesters were arrested. The LCITF subsequently constructed a truck wash pad inside the gates where tires were hosed off before every trip out of the site, and the wash water was collected for safe disposal.

In December 1978, the DEC announced that lateral trenching across the canal site was necessary because only a limited amount of leachate was entering the north-south trenches. Residents feared lateral trenching would trigger explosions and release toxic vapors and pressed their demands for additional relocation in public protests. On February 8, 1979, temporary relocation at the state's expense in local motels was made available for pregnant women and children under two living between Ninety-seventh and 103rd Streets until the child was two. Relocation was refused for those planning pregnancy. On August 24, a second temporary relocation was allowed for those with illnesses related to fumes from the remedial construction. Temporary relocation was stressful and exhausting. Many families were split between rooms, and most had mother stay at the motel with the children while father stayed at home to protect the property. (Right, Dunmeir family; below, Ed Posniak.)

The grassroots LCHA was resource-poor, especially in comparison with the other established newsmakers at Love Canal (Hooker Chemical and government agencies). Lois Gibbs and the LCHA quickly recognized the powerful role positive media coverage could play in their quest for additional permanent relocation of outer ring residents from Ninety-third to 103rd Streets and began a campaign of protests and events that captured the media's attention. The LCHA would become the most cited print news source in local papers and the *New York Times*. Lois Gibbs, Debbie Cerrillo, and other members of the LCHA appeared on local radio and television talk shows and appeared on the *Phil Donohue Show* (October 1978 and June 1980) and *Good Morning America* (September 1980). (Above, Debbie Cerrillo; below, Lois Gibbs.)

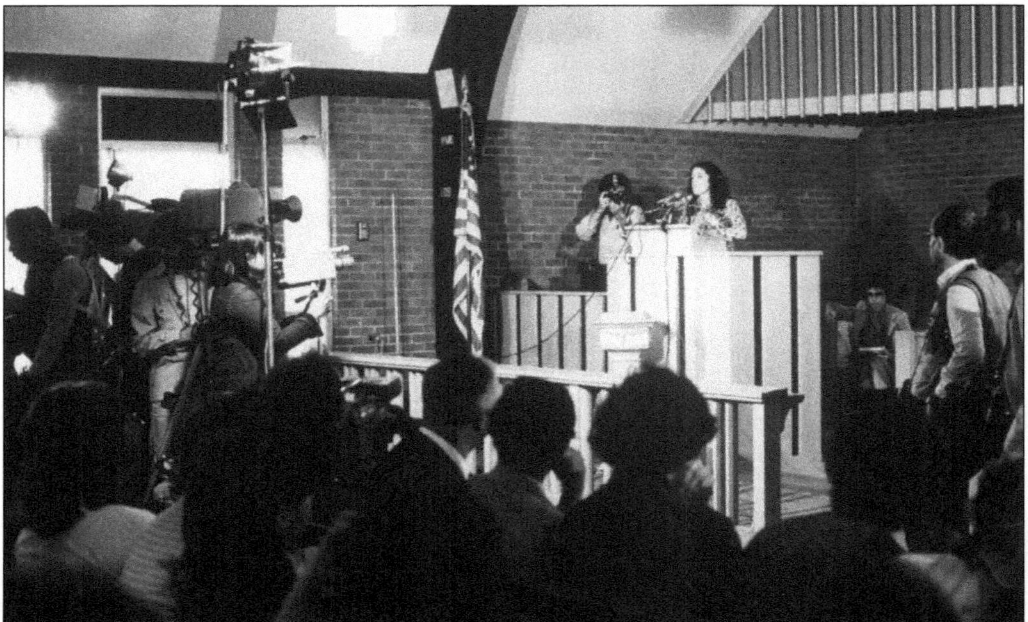

In May 1979, the state held an auction for the 239 Rings 1 and 2 homes purchased in August 1978. Fifteen bids were received. The LCHA picketed the auction, and a number of neighboring towns announced that Love Canal houses would not be allowed in. On May 29, remedial construction began on the center and northern sections of the site and the permanent treatment facility. In June 1979, President Carter announced plans for a $1.63-billion Superfund for the cleanup of other hazardous waste sites across the country. In September 1979, the state announced that it would demolish all Ring 1 houses. By this time, 425 residents had been temporarily relocated.

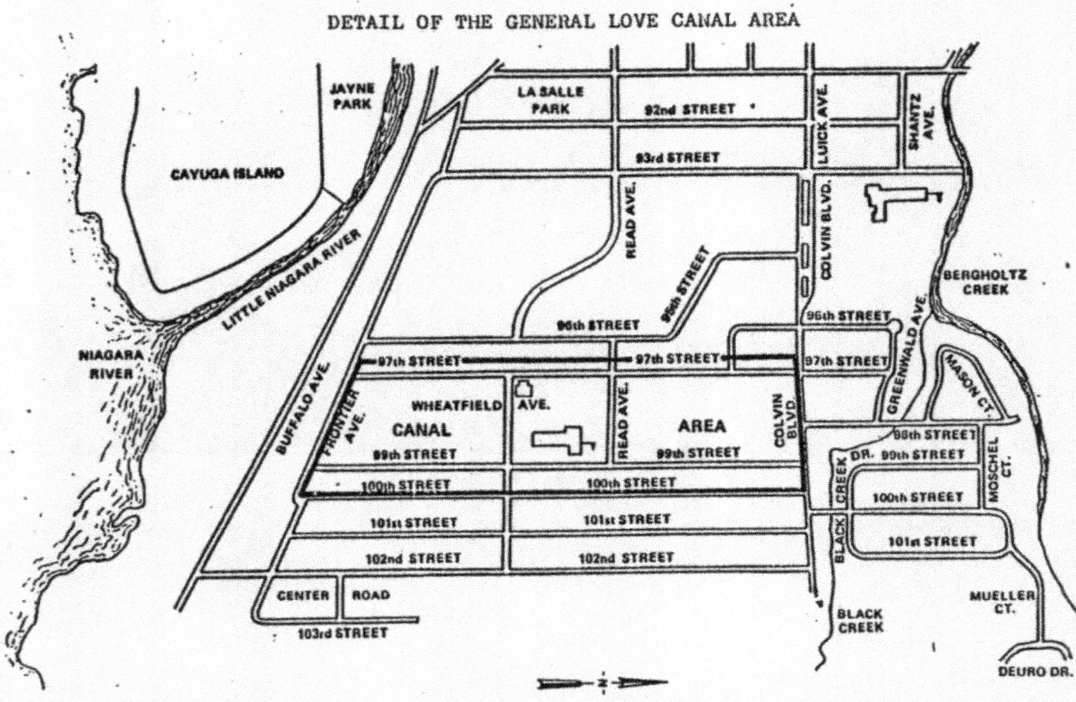

By October 1979, remediation was completed on the southern section of the site, and those relocated were told to return home unless they had medically documented illnesses. Residents refused to leave the motels and sought a court order requiring continued relocation. On October 16, Governor Carey announced that additional homes would be purchased. On October 26, Carey announced the second permanent relocation of all residents wanting to leave the area and the formation of the Love Canal Area Revitalization Agency (LCARA), a task force charged with coordinating the home purchases and stabilizing and revitalizing the area. The remediation was completed in November, and the 110 relocated families were told to return home, but they refused. Dioxin was detected in Bergholtz Creek north of the declared area on November 10, 1979. (DOH.)

On May 17, 1980, the EPA released data indicating chromosome damage in residents that might require relocation of 710 residents from the surrounding area. On May 19, EPA officials came to the LCHA headquarters on Colvin Avenue to discuss the findings with Lois Gibbs. A crowd of residents began to gather outside, and Gibbs, referring to the hostage crisis in Iran, told the officials that they appeared to be hostages of the angry crowd outside. On May 22, President Carter declared the area a federal emergency for the second time, and the area was extended from Ninety-third to 103rd Streets and from Frontier Avenue to Bergholtz Creek.

A third temporary relocation began on May 23, 1980. Many homeowners from the outer rings, including Marie and Ed Posniak, who evacuated their home on Colvin Avenue, chose to leave their homes.

On October 1, 1980, at ceremonies at the Niagara Falls airport, President Carter and Governor Carey (shown in these photographs) publicly signed joint agreements for the federal government to provide $15 million to the state. A total of $7.5 million of it was to pay for the relocation of residents choosing to leave the area. The LCARA was named the administrator of the relocation fund and charged with renovating and maintaining the houses.

Abandoned Ring 1 homes on Ninety-ninth Street await demolition, as seen from Wheatfield Avenue in May 1982.

Seven

Neighborhood Revitalization 1981–2012

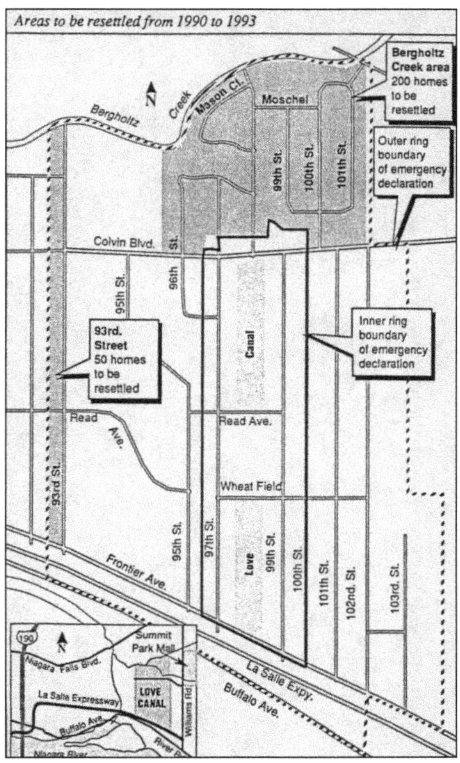

By March 1981, out of the eligible 550 homes in the expanded emergency declaration area, 228 had been sold to the LCARA; by November, 386 had been sold. As of June 1981, the cleanup costs of Love Canal included $14.8 million spent by the state, $21 million by the federal government, and $6.3 million by the City of Niagara Falls. (SUNY.)

By March 1982, the state DEC had collected over 100,000 gallons of liquid chemical waste and accumulated 20,000 tons of thick leachate in 450 barrels stored on site. In 1984, the barrels of leachate were buried under the clay cap. (SUNY.)

Ring 1 and 2 homes had been boarded up and abandoned since the first permanent relocation in August 1978, and most were in disrepair. In January 1982, the state announced that Ring 1 and 2 houses would be demolished. Here are some homes as they were in May 1982.

Here is a view of abandoned Ninety-ninth Street homes as they were in May 1982, just before they were demolished.

Ninety-ninth Street School is visible in the background, and the permanent treatment facility is on the left as seen from Ninety-ninth Street in May 1982.

As seen from Frontier Avenue in May 1982, the Ninety-ninth Street School is in the distance with the clay capping extending to the backs of houses on the right.

In June 1982, the demolition of the 237 abandoned Rings 1 and 2 homes began.

Ceiling and wall paint that had been exposed to the ambient air inside a boarded-up Ring 1 home had begun blistering and peeling after three years of exposure to apparently contaminated air, as seen in these May 1982 photographs.

In June 1982, Ring 1 and 2 homes were bulldozed, one after another, and the rubble was buried within their foundations.

Ninety-ninth Street homes are seen being bulldozed and buried in June 1982. The Ninety-ninth Street School is visible in the background. It was demolished on June 8, 1983, and the Ninety-third Street School was demolished on May 31, 2000.

All the Ninety-ninth Street homes were bulldozed and buried up to the Wheatfield Avenue intersection in June 1982. The rest of the homes were demolished by the end of July 1982.

An abandoned protest sign, painted on what was a for-sale sign, lies in a pile of demolition debris. This sign, with two messages, would become part of the rubble pile that was bulldozed, buried, and entombed in what had been the basement of someone's former dream house.

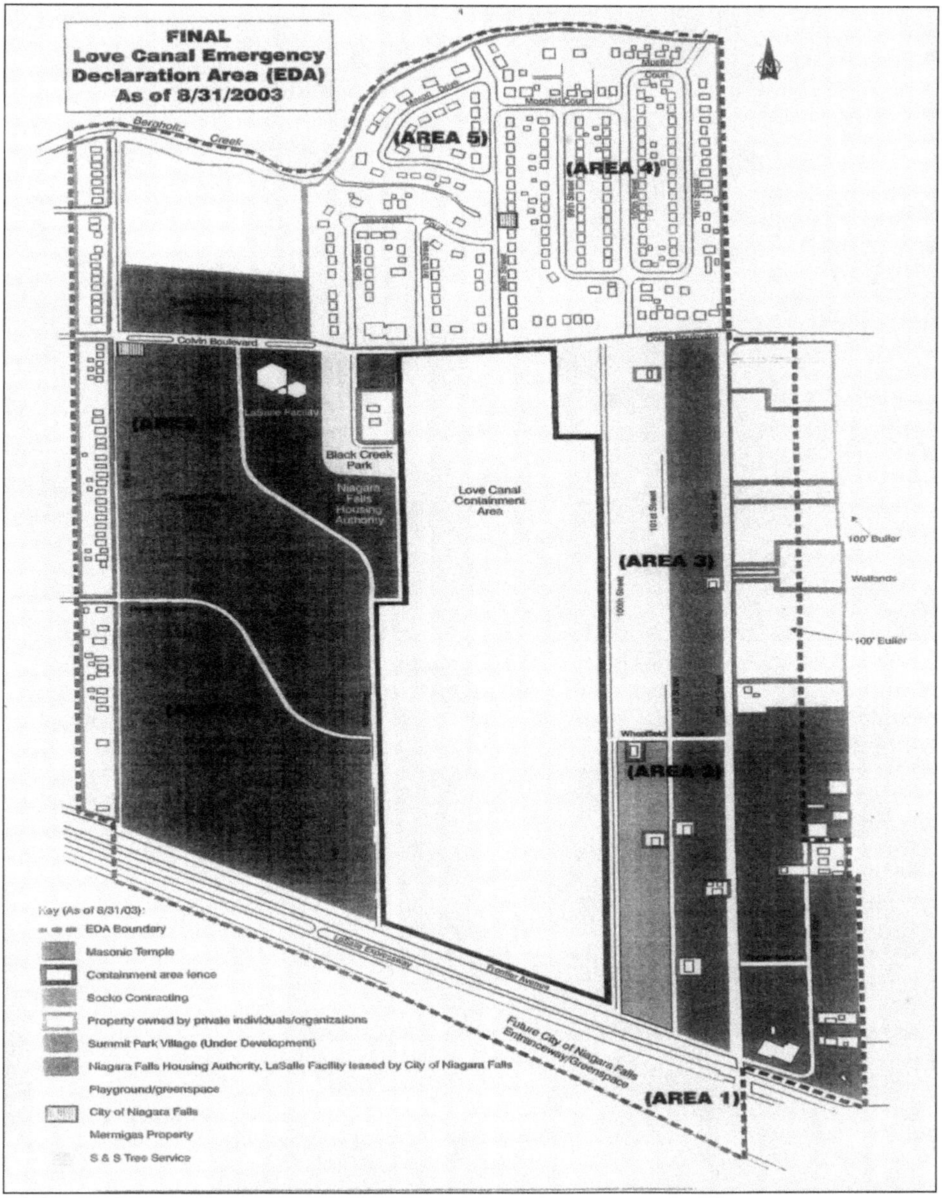

On July 15, 1982, the EPA announced that the cleanup and containment projects had worked, and except for Rings 1 and 2, the neighborhood was again habitable. The EPA recommended that all storm sewers and creek beds in the outer areas be scoured before resettlement, and on May 2, 1988, the cleanup of Bergholtz and Black Creeks began. On September 27, 1988, the DOH announced its habitability decisions, declaring Areas 2 and 3 south of Colvin Boulevard non-habitable and Areas 4, 5, 6, and 7 as habitable. In June 1990, Areas 2 and 3 were designated "commercial light industrial use," and Areas 4, 5, 6, and 7 were designated as "residential use." Areas 4 and 5 were renamed Black Creek Village; 239 homes with $12 million in property value were restored and reoccupied by May 2000. On September 30, 2004, Love Canal was removed from the Superfund list that it gave rise to. The EPA declared the cleanup and containment work complete—more than two decades after 950 families were evacuated. (LCARA.)

As seen from Colvin Avenue looking north in September 2011, this view is of the revitalized 100th Street in the newly named Black Creek Village, just across Colvin Avenue from the fenced-off Love Canal containment area. Coming full circle, these revitalized houses became dream homes for first-time buyers or retirees.

As seen from Colvin Avenue looking south in September 2011, this is a view of the remains of Ninety-ninth Street, with the containment area fence and former dump site on the right and ending at the fence in the background.

As seen from Colvin Avenue, here is a September 2011 view of 101st Street with a house occupying the street. At that time, three houses still existed on 101st Street. Also, 102nd and 103rd Streets had one house each.

The Love Canal containment area is seen through the fence from Frontier Avenue. Behind the fence is what looks like a wide, grassy meadow, under which are entombed Love's dream canal, Hooker's dump site, and the rubble of hundreds of American dreams.

This is a view of 100th Street as seen from Frontier Avenue with garbage on the side of the road in September 2011. Ironically, despite being part of the most infamous dump site in history, garbage (tires, televisions, trees, and yard waste) had been dumped on the right side of this abandoned street.

On August 7, 2003, the 25th anniversary of the Love Canal emergency declaration, a monument was erected on Ninety-third Street. Engraved on it is a chronology of Love Canal events. Frank Cornell, the last executive director of the LCARA, designed and paid for it.

Eight

Lessons of Love Canal
The Future

Love Canal became the infamous icon and harbinger of the national and international hazardous waste disposal crisis. Love Canal was instrumental in the passage of federal Superfund legislation and the need for cradle-to-grave legislation to control hazardous substances. The total cost of Love Canal is estimated at more than $400 million. More than $11 billion in lawsuits were filed in response to Love Canal. In 1983, a lawsuit filed by 1,328 residents against Occidental Chemical Corporation (Hooker's name was changed Occidental Chemical in 1989) was settled for $20 million, of which $1 million was set-aside for a medical trust fund. In 1994, Occidental agreed to pay the state $98 million in settlement, and in 1995, it settled with the federal government for $129 million to cover cleanup costs. The LCHA would become the model for many other grassroots citizens' organizations seeking environmental and social justice for their neighborhoods and is also credited with reinvigorating the environmental movement, inspiring the environmental health movement, and extending NIMBY ("Not In My Backyard") to NIABY ("Not In Anyone's Backyard").

This is a view of 101st Street as seen from Colvin Avenue in September 2011. The Love Canal neighborhood no longer exists. All that remains are the skeletons of streets, sidewalks, and driveways leading nowhere; street signs directing no one; and a few houses occupied by those who never felt fear. There is a prevailing sense of desolation mixed with resolution, and as the pavement crumbles and the wildness of nature encroaches, the area appears to be returning to the earth.

Bibliography

Adams, Edward Dean. *Niagara Power: History of the Niagara Falls Power Company 1886–1918: Volumes I & II*. Niagara Falls, NY: Niagara Falls Power Company, 1927.

Blum, Elizabeth D. *Love Canal Revisited: Race, Class, and Gender in Environmental Activism*. Lawrence: University Press of Kansas, 2008.

Breton, Pierre. *Niagara: A History of the Falls*. Albany: State University of New York Press, 1982.

Brown, Michael. *Laying Waste: The Poisoning of America by Toxic Chemicals*. New York City: Pantheon Books, 1979.

Dumych, Daniel M. Images of America: *Niagara Falls, Volume I*. Charleston, SC: Arcadia Publishing, 1996.

———. Images of America: *Niagara Falls, Volume II*. Charleson, SC: Arcadia Publishing, 1998.

Gibbs, Lois. *Love Canal: My Story*. Albany: State University of New York Press, 1982.

———. *Love Canal: The Story Continues*. Gabriola Island, BC: New Society Publishers, 1998.

———. *Love Canal and the Birth of the Environmental Health Movement*. Washington, DC: Island Press, 2011.

New York State Department of Health. *Love Canal—Public Health Time Bomb: A Special Report to the Governor and Legislature*. Albany: September 1978.

———. *Love Canal: Special Report to the Governor & Legislature*. Albany: April 1981.

Ploughman, Penelope D. *The Creation of Newsworthy Events: An Analysis of Newspaper Coverage of the Man-Made Disaster at Love Canal*. SUNY Buffalo: doctorate dissertation, 1984.

———. "Disasters, the Media and Social Structures: A Typology of Credibility Hierarchy Persistence Based on Newspaper Coverage of the Love Canal and Six Other Disasters." *Disasters* 21 (June 1997): 118–137.

———. "The Local Newspaper as Initiator, Definer, Diffuser, and Legitimator of Community Controversy: The Role of the *Niagara Gazette* in the Love Canal Hazardous Waste Landfill Disaster." *Newspaper Research Journal* 16 (Spring 1995): 56–75.

———. "The Print News Media 'Construction' of Six Natural Disasters." *Disasters* 19 (December 1994): 308–326.

Thomas, Robert E. *Salt & Water, Power & People: A Short History of Hooker Electrochemical Company*. Niagara Falls, NY: Hooker Electrochemical Company, 1955.

www.ingramcontent.com/pod-product-compliance
Lightning Source LLC
LaVergne TN
LVHW070819030926
847199LV00039B/506
9780738575605